AF439745

The WBF Book Series: Volume 1

ISA-88 Implementation Experiences

The WBF Book Series: Volume 1

ISA-88 Implementation Experiences

By WBF

Edited by
William Hawkins
Dennis Brandl
Walt Boyes

Momentum Press, LLC, New York

ISA-88 Implementation Experiences
Copyright © Momentum Press®, LLC, 2010

First published in 2010 by
Momentum Press®, LLC
222 East 46th Street, New York, N.Y. 10017
www.momentumpress.net

ISBN-13 (hard back, case bound): 978-1-60650-212-9
ISBN-10 (hard back, case bound): 1-60650-212-3

ISBN-13 (e-book): 978-1-60650-214-3
ISBN-10 (e-book): 1-60650-214-X

DOI forthcoming

Volume 1 in the WBF Book Series by WBF, edited by William Hawkins, Dennis Brandl, and Walt Boyes, published by Momentum Press®, LLC

Cover Design by Jonathan Pennell
Interior Design by Scribe, Inc. (www.scribenet.com)

First Edition: December 2010

10 9 8 7 6 5 4 3 2 1

Printed in Taiwan

Contents

Figures

Tables

WBF Foreword

The purpose of this series of books from WBF, The Organization for Production Technology, is to publish papers that were given at WBF conferences so that a wider audience may benefit from them.

The chapters in this series are based on projects that have used worldwide standards—especially ISA-88 and 95—to reduce product variability, increase production throughput, reduce operator errors, and simplify automation projects. In this series, you will find the best practices for design, implementation, and operation and the pitfalls to avoid. The chapters cover large and small projects in a wide variety of industries.

The chapters are a collection of many of the best papers presented at the North American and European WBF conferences. They are selected from hundreds of papers that have been presented since 2003. They contain information that is relevant to manufacturing companies that are trying to improve their productivity and remain competitive in the now highly competitive world markets. Companies that have applied these lessons have learned the value of training their technical staff in relevant ISA standards, and this series provides a valuable addition to that training.

The World Batch Forum was created in 1993 as a way to start the public education process for the ISA-88 batch control standard. The first forum was held in Phoenix, Arizona, in March of 1994. The next few years saw growth and the ability to support the annual conference sessions with sponsors and fees.

The real benefit of these conference sessions was the opportunity to network and talk about or around problems shared by others. Papers presented at the conferences were reviewed for original technical content and lack of commercialism. Members could not leave without learning something new, possibly from a field thought to be unrelated to their work. This series is the opportunity for anyone unable to attend the conferences to participate in the information-sharing network and learn from the experiences of others.

ISA-88 was finally published in 1995 as *ISA-88.01-1995 Batch Control Part 1: Models and Terminology*. That same year, partially due to discussions at the WBF conference, ISA chartered ISA-95 to counter the idea that business people should be able to give commands to manufacturing equipment. The concern was that business people had no training in the safe operation of the equipment, so boardroom control of a plant's fuel oil valve was really not a good idea. There were enough CEOs smitten with the idea of "lights-out" factories to make a firewall between business and manufacturing necessary. At the time, there was a gap between business computers and the computers that had infiltrated manufacturing control systems. There was no standard for communication, so ISA-95 set out to fill that need.

As ISA-95 began to firm up, interest in ISA-88 began to wane. Batch control vendors made large investments in designing control systems that incorporated the models, terminology, and practices set forth in ISA-88.01 and were ready to move on. ISA-95 had the attention of vendors and users at high levels (project-funding levels), so the World Batch Forum began de-emphasizing batch control and emphasizing manufacturing automation capabilities in general. This was the beginning of the transformation of WBF into "The Organization for Production Technology." Production technology includes batch control.

The WBF logo included the letters "WBF" on a map of the world, and since this well-known image was trademarked, the organization dropped the small words "World Batch Forum" entirely from the logo after the 2004 conference in Europe. WBF is no longer an acronym. Conferences continued annually until the economic crash of 2008. There was no conference in 2009 because many companies, including WBF, were conserving their resources.

WBF remained active and solvent despite the recession, so a successful conference was held in 2010 using facilities at the University of Texas in Austin. Several papers spoke of the need for procedural control for continuous and discrete processes. The formation of a new ISA standards committee (ISA 106) to address this need was announced as well. Batch control is not normally associated with such processes, but ISA-88 has a large section on the design of procedural control. There is a need for a way to apply that knowledge to continuous and discrete processes, and some of those discussions will no doubt be held at WBF conferences, especially if the economy recovers. We would like to invite you to attend our conference and participate in those discussions.

WBF has always been an organization with an interest in production technologies beyond batch processing, even when it was officially "World Batch Forum." Over the years, as user interests changed, so has WBF. We have not lost our focus on batch; we have widened our view to include other related technologies such as procedural automation. We hope you will find these volumes useful and applicable to your needs, whatever type of process you have, and if you would like more information about WBF, we are only a simple click away at http://www.wbf.org.

William D. Wray, Chairman, WBF
Dennis L. Brandl, Program Chair, WBF
August 2010

Foreword by Walt Boyes

Many years ago, some dedicated visionaries realized that procedure-controlled automation would be able to codify and regularize the principles of batch processing. They set out on a journey that eventually arrived at the publication of the batch control standard ISA-88 and the development of the manufacturing language standard ISA-95.

Many end users have benefited from the work of these visionaries, who founded not only the ISA-88 Standard Committee but also the WBF. WBF has been an unsung hero in the conversion of manufacturing- to standards-based systems.

Today, WBF continues as the voice of procedure-controlled automation in the process, hybrid, and batch processing industries. The chapters that make up this book series provide a clear indication of the power and knowledge of the members of WBF.

I have been proud to be associated with this group of visionaries for many years. *Control* magazine and ControlGlobal.com are and will continue to be supporters of WBF and its aims and activities.

I would like to invite you to come and participate in WBF both online and at the WBF conferences in North America and Europe that are held annually. You will be glad you did. You can get more information at http://www.wbf.org.

Walt Boyes, ISA Fellow
Editor in Chief
Control magazine and ControlGlobal.com

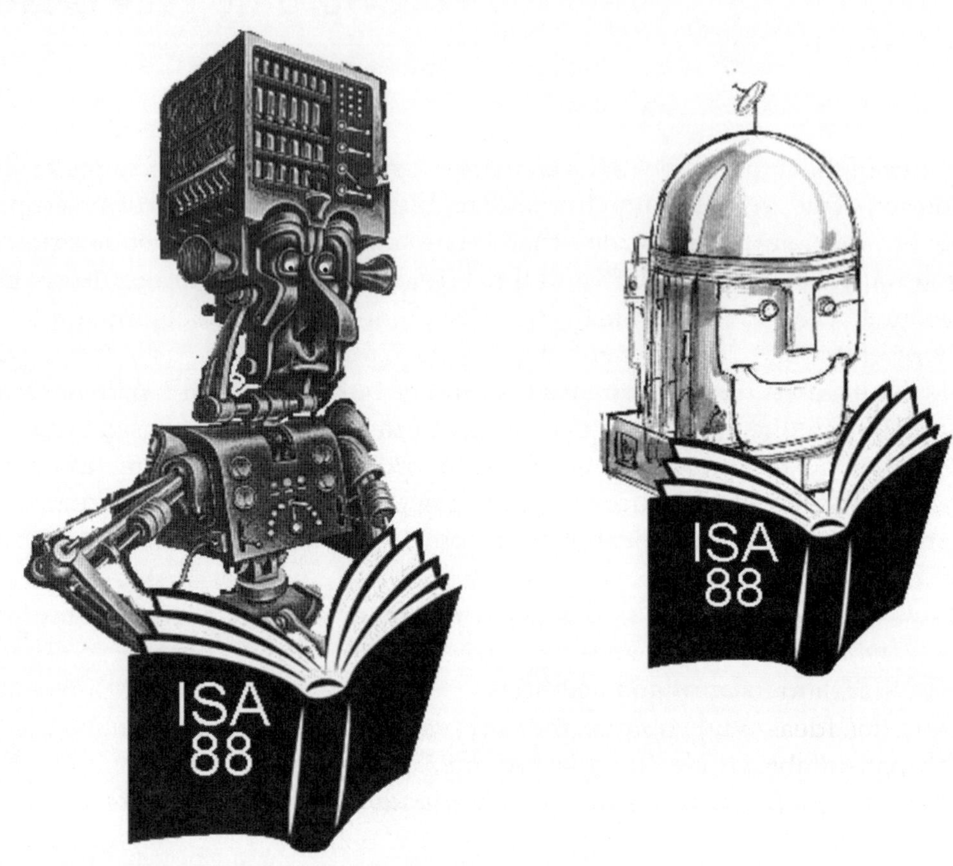

Automation engineers old and new need to read ISA-88.

Preface

Twenty-one of the chapters in this book were selected from WBF conference papers because they were related to the implementation of batch control in one way or another. In addition, the editor has added a twenty-second chapter on material transfers that was not previously presented at a WBF conference because several of the other chapters demonstrate the need for a standard on models and terminology for transfers.

It is the need to use recipes to create products that complicates batch process design. In the bad old days (or good if you were a custom systems designer), there was no consistent set of models and terminology for turning recipes into processes. Vendors and users evolved their own ways of describing batch process control. This gave way to a "Tower of Babel" situation. The need for standard models and terminology grew with each new system and each new generation of designers.

The ISA Standards and Practices Board created ISA88 in 1988. A small (and changing) group of ISA committee members hammered out, not to say word-smithed, a standard that was originally based on Tom Fisher's 1990 ISA book *Batch Process Control*. Ideas were submitted, reviewed, discussed, refined, and finally put into a format suitable for an international standard.

Often, the proponent of a new idea had to educate the group so that they would understand why it was a good idea. The separation of recipe and equipment procedures was one of those ideas. It eliminated specific hardware references in recipe procedures, at a time when nearly everyone was used to putting hardware references into their recipes. The drawback of this methodology was that the recipe had to be changed if the equipment was improved or replaced. Generally, there are lots more recipes than pieces of equipment, and recipes have to be revalidated for food and pharmaceutical processes. This made recipe systems expensive to implement and maintain. The new model eliminated all of these problems.

Many outside the committee were unaware of the usefulness of the new ideas. Some were aware but did not fully understand how to use them. Some went ahead and built systems anyway, claiming compliance with the standard. The WBF was formed to address the education problem by providing a forum for control experts to present and share their knowledge and demonstrate, through real projects, the advantages of using the ISA-88 models.

The name of the standard evolved over the years. These chapters have been edited to use "ISA-88.01" when referring to part 1 of the standard for batch process control that was published by the ISA as *ISA-88.01-1995 Batch Control Part 1: Models and Terminology.* "ISA-88" may be used to refer to all of the parts of the standard, or it may be used to mean the general concepts of the standard without getting specific.

The biggest misunderstanding of all is that ISA-88.01 only applies to automation engineers. It actually lays out the design process for recipe-driven manufacturing. ISA-88.01 should be applied before any decisions are made about which processes can be automated and which processes require human operators who are closer to understanding the process than computer programmers. See Lynn Craig's fine discussion of this subject in Chapter 4.

ISA-88.01 represented a major development in process control, and it meant change for people who had different meanings for the terms the committee chose. While using the new models and terminology was painful for some, it provided major benefits for all manufacturing companies. Without common terminology and models there was confusion in process companies as they tried to describe their production processes to vendors and as vendors tried to define their solutions to process owners. With the ISA-88 models and terminology, confusion was reduced, process owners could describe their requirements in terms vendors could understand, process owners could easily compare and evaluate different vendor solutions, and vendors could explain their solutions with less misunderstanding.

Some wrongly believe that the ISA-88 standard only applies to the process industries and batch processes. While that was the initial focus of the ISA-88 committee, the resultant work has been effectively applied in nonprocess industries and discrete and continuous processes. The ISA-88 standard can be applied to any process where procedural control is needed. While this is most common in batch processes, it is also used in startup, shutdown, exception handling, and switch-overs in all manufacturing processes.

The ISA-88 standard is a dictionary of terms and concepts. The chapters in this book describe projects that converted the ISA concepts into reality. These chapters represent a common body of knowledge about how to apply the ISA-88 standard in all aspects of a project—from concept to design, implementation, maintenance, and upgrades.

Happy reading and remember: when in doubt, refer to *ISA-88.01-1995 (R2006) Batch Control Part 1: Models and Terminology* (or ask ISA for the latest version). The European version is IEC 61512-1, published in 1997.

Bill Hawkins
August 2010
wmh@iaxs.net

How a Flexible Batch Application Can Save Costs

Presented at the WBF
North American Conference,
March 24–26, 2008, by

J. Gordon Roney
Control Systems Group Leader
groney@norus.jnj.com
Noramco, Inc., 1440 Olympic Drive,
Athens, GA 30601, USA

Andrew Blankenship
Integration Engineer
awb@innovativecontrols.com
Innovative Controls, Inc.,
624 Reliability Circle, Knoxville, TN 37932, USA

Abstract

This chapter is a case study of the implementation of batch-capable Equipment Modules (EMs) at an existing facility that manufactures Active Pharmaceutical Ingredients (API) that can be used with or without a batch manager application.

In 2004, we started the process of expanding a recently installed process control system. We desired the ability to reuse EMs in order to save costs on system configuration, qualification, and new recipe configuration. Additionally, we wanted the capability to use these EMs independent of the batch manager application.

The EMs resided in the controllers instead of at the batch management level, which allowed us to configure and qualify them independent of any higher-level recipe manager. Cost savings resulted from being able to qualify them once and use them in multiple instances and recipes. The cost of configuring new recipes was also lowered because the EMs were already installed and qualified, which reduced the time required to write and qualify the new recipe.

Monitoring of process upset conditions and the response to these conditions is implemented directly at the EM instance instead of through the batch application manager.

This chapter will illustrate the following:

- Configuration of independent EMs

- Cost-savings calculation

- Integration of process monitoring at the EM without the batch management system

- Project management lessons learned

Introduction

Noramco, Inc. manufactures active pharmaceutical ingredients, narcotics, and medical devices in three production facilities at its Athens, Georgia, site. The oldest of the production facilities, Building 1, is used to produce active pharmaceutical ingredients. Historically, the production processes carried out there have been manual operations. In 2002, to meet changing business needs and planned expansions, Noramco began the process of installing a process control system for use in Building 1.

As with all projects, requirements are a hard fact. The planned installation of the process control system was no different. The three main requirements for the implementation of the process control system were the following:

1. The operation of the production process, via the process control system, should support both direct operator interaction and a batch application manager when installed.

2. EMs and EM control strategies that have already been qualified and implemented should not require a requalification when the batch application manager is installed.

3. The automation strategy should support the automation of new units, reusing as much existing work as possible.

The decision to implement an ISA-88.01 based control strategy for the Building 1 process control system was not in question due to the following three factors:

1. Noramco's previous success in implementing an ISA-88.01 based control system onsite
2. The overriding requirements of the process control system
3. The strictly batch nature of production in Building 1

But the ISA-88.01 model that best addressed the requirements was yet to be decided.

Further evaluation of the requirements resulted in the following EM control options: direct control by the operator, control via a higher-level operation, and control via the batch application manager. In all of the EM control options listed, the ability to select the needed EM, the appropriate EM control strategy, and the input parameters for the control strategy and a means to monitor and control the EM once execution was under way was needed. Since the batch application manager was not going to be implemented in the first phase of the process control system installation, the ability for the operator to control the EM independent of the batch application manager clearly became the driving force in determining the appropriate ISA-88.01 model.

Software Engineering Design and Implementation

Various control recipe procedure-to-equipment control relationships were evaluated to determine the best way to fulfill these requirements. The result of this evaluation was the selection of a collapsed ISA-88.01 model (Fig. 1.1).

The collapsed model allowed equipment control independent of the batch application manager. This independence was accomplished by placing equipment phases—the right side of the collapsed ISA-88.01 model—in the controllers of the process control system. The recipe procedure and phases—the left side of the collapsed ISA-88.01 model—would be implemented in the batch application manager.

Design and implementation of standard EMs and subordinate Control Modules (CMs) were made much easier by making extensive use of object-oriented programming. Object-oriented programming uses "classes" and "objects" to develop computer-based applications. Classes define the abstract idea, and objects are the real-world application of the class. It is often easier to think of a "class" as a model or template—in this case an EM template. For example, a number of units may require an EM with the same basic functionality (e.g., an agitator with speed control). In order to promote reuse, an EM template named "Agitator with Speed

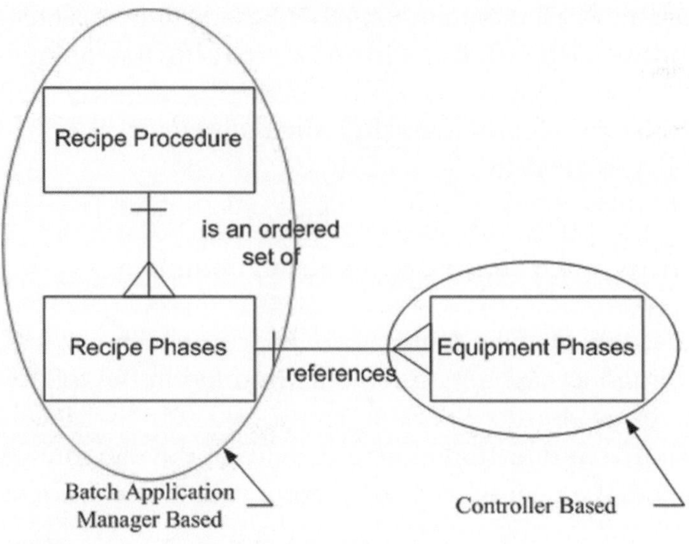

Figure 1.1.   The collapsed ISA-88.01 model selected.

Control" would be developed. This template would not have links to subordinate CMs. Each individual application based on the EM template would have the links to subordinate CMs.

The first major implementation activity was the introduction of CM templates. By defining, developing, and qualifying CM templates, the links needed by EMs would be predetermined and consistent (Fig. 1.2).

The next major implementation activity was the introduction of EM templates. Creating a new EM template is a multiple-step process. The first step is

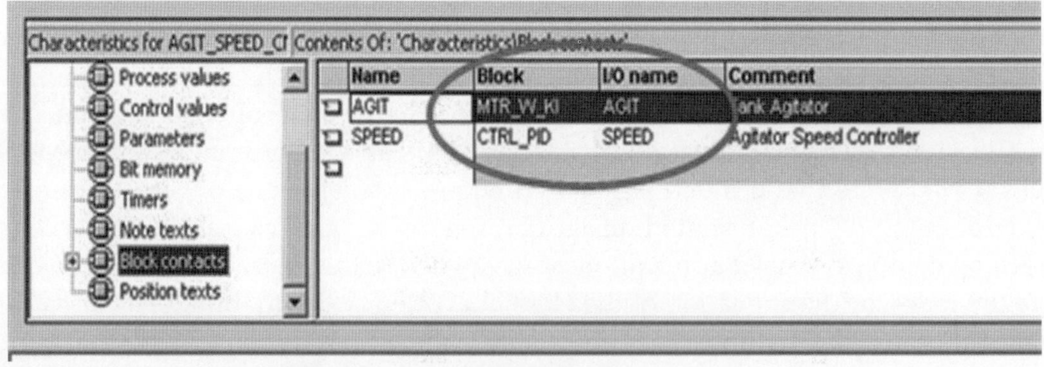

Figure 1.2.   CM template usage within an EM template.

development of the EM template functional specification. This document defines how the EM and its phases will function. The second step is the implementation of the EM template. Once an EM template had been developed, the first instance was created followed by the qualification of the first instance and its EM template.

Figure 1.3 is a snapshot of the EM phases for the Agitator with Speed Control EM class. Depending on which phase is executed, the agitator may be started or stopped and the function of the speed controller would be set (e.g., control at setpoint, manual with a fixed output, and so forth).

When an application requires the use of an existing EM, an instance of the qualified EM template is created. Creating the tangible EM from the EM template is known as "instantiation." Figure 1.4 depicts an instance of the Agitator with Speed Control EM. Circled on the top left side of the figure is the specific instance name, on top, with the EM template name below it. Connection points for the generic CM names defined in the EM to the subordinate CMs themselves are circled below this, on the left side. In this example, the agitator speed setpoint "SP_AI" (circled on the left side) is of one of the parameters an operator or batch application manager can change as part of the available phases. The links to the subordinate CM (in this case the agitator speed controller) are circled on the bottom right of Figure 1.4.

Because an instance is an exact functional replica of an EM template that has already been qualified, it is not necessary to fully qualify each instance's functionality. Only the instance's unique attribute values must be qualified. In practice, this is equivalent to the EM instance interaction with subordinate CMs and any parameters that can be modified for that instance (e.g., setpoints). Once in service, implementation of changes is simplified because instances inherit the functionality of the EM

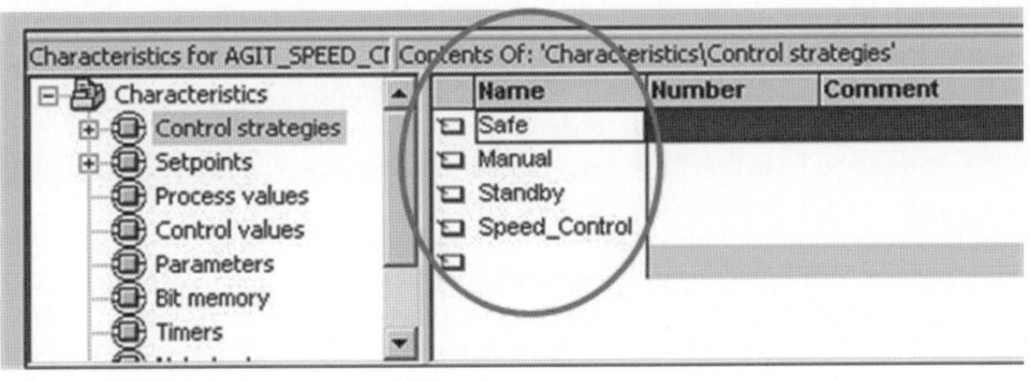

Figure 1.3.   EM phases.

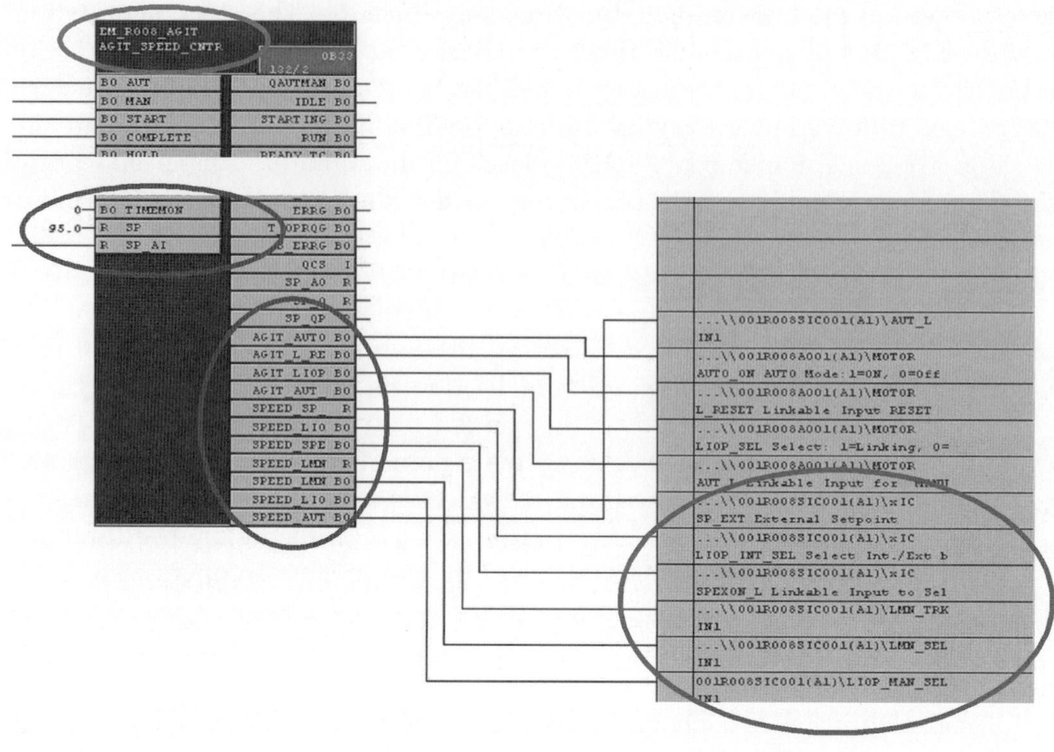

Figure 1.4. Agitator with Speed Control instance.

template. A change is implemented and tested once at the EM template, and the functionality is propagated to each instance of the EM template.

Another important aspect of implementing EMs at the controller level is related to the safe, normal operation of equipment. The response of the EM to an initiating event or events that warrant the subordinate CMs to assume a predesignated state can be implemented and tested prior to the implementation of the batch application manager. It is important to note that this functionality does not serve the functionality of a safety-instrumented system.

Most if not all EMs that have been implemented to date have a dedicated phase called "safe" for this predetermined state. Initiation of this phase can be internal to the EM, via operator initiation of the "safe" phase, or initiated externally to the EM itself (Fig. 1.5). Examples of these external events include the shutdown of a unit, power loss, cascaded unit shutdowns, and so on. When the "safe" phase is called upon to execute, the EM will execute the predetermined actions. By monitoring and responding to initiating events at the controller level directly, the response time can be reduced as compared to EMs hosted on a server-based batch application manager.

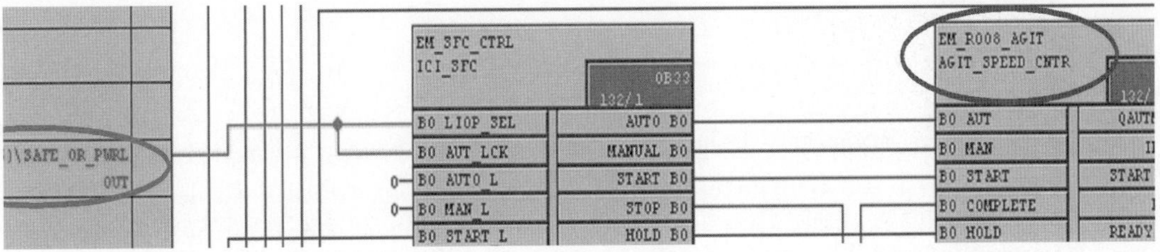

Figure 1.5.   EM "safe" implementation.

Cost Savings

Pharmaceutical manufacturers, like most other manufacturing companies, undertake the standard activities of engineering configuration, preimplementation testing, and operational testing when implementing control systems and new control strategies. Unlike other manufacturing companies that do not operate in an FDA-regulated environment, pharmaceutical manufacturers must undertake the following additional activities when implementing control systems and new control strategies:

- Developing documentation to describe what the control system and its control strategies should do (user requirements)
- Defining and documenting how the task is to be done (functional specifications)
- Documenting the qualification and commissioning rationale (e.g., impact assessments, project qualification plans)
- Testing (e.g., software acceptance, installation, and operational qualification tests)

The extensive hours and expenses associated with these tasks make the reuse of EMs, support documentation, and testing extremely appealing.

It should come as no surprise that the largest investment of time, and therefore expense, comes in the development of the first instance of the EM. Once the first instance is created and qualified, subsequent instances do not need to be functionally qualified to the extent of the original model. In contrast, if no EM templates are used, then complete testing of functionality must be undertaken because no defined template has been qualified. Two case studies comparing the initial EM implementation to a subsequent implementation of the same type are detailed in the following sections. These examples demonstrate the cost savings afforded by controller-based EMs.

Case Study One

The first case study involves the implementation of an Agitator with Speed Control EM. The original implementation occurred during 2004 and the second implementation in 2007. The EM consists of an agitator and an agitator speed controller. Four relatively simple control strategies are available. Table 1.1 lists the time for each major implementation task and the time savings gained for the second implementation.

The overall time savings for the second implementation was approximately 86 hours, or a 45% reduction in labor hours. The greatest percentage decrease occurred in the task related to requirement and specification development, which took 67% less time for the second implementation. This large drop can be attributed to the copying or revision of existing documentation. In terms of hours saved, engineering configuration tasks experienced the greatest drop of 36 hours, or approximately 50%. In this case, the time savings can clearly be attributed to the reuse of the EM. The hours presented for both the original implementation and second implementation include the cost for implementing the underlying CMs.

Case Study Two

The second case study involves the implementation of a tank temperature control EM. The original implementation occurred during 2004 and the second implementation in 2005. The purpose of this EM is to control the equipment used for heating, cooling, and maintaining the temperatures of tanks and reactors. The EM consists of a pump, multiple valves, a tank temperature controller, and a jacket temperature controller. Six control strategies are available. The EM phases sequence the starting and stopping of the pump, the opening and closing of valves, the configuration of

Table 1.1. Case study one: Time savings for implementation tasks, in hours			
Task	*Original*	*Second*	*Saved*
Requirements and specifications	24	8	16
Qualification documentation	80	51	29
Engineering configuration	73	37	36
Preimplementation testing	12	8	4
Total	189	104	85

temperature control loops, the ramping of temperatures, the monitoring of conditions, and so forth. In this case study the time for requirement, specification, and qualification documentation development is not included, since the original implementation was part of a larger project and could not be easily broken out (Table 1.2).

As in case study one, the hours presented for both implementations include the cost for implementing the underlying CMs. In this case study, the task engineering configuration experienced a drop of 55%. The drop of 55% is in line with the percentage reduction in hours for the same task in case study one (by 50%). Preimplementation testing for the second implementation took 50% less time than the original implementation, due to the reduced testing of complex EM logic. Although not presented here, a large time savings was noted during the second implementation for both requirement and specification development and qualification documentation development.

Implementation of a batch application manager began in mid-2005 and hit full stride during 2007. No requalification of the underlying EMs has been required because the batch application manager uses the same EM attributes to execute the EM, as does an operator. During the qualification of recipes, verification that the correct parameters and that the correct phase is executed at the EM level is performed. The states of the CMs within the EM are not verified because they were verified in previous testing. This has resulted in less complex and faster qualification of recipes, compared to solely recipe-driven production processes on-site.

Lessons Learned and Additional Advantages

Outside of the obvious time and cost savings gained via the reuse of EMs and the operational benefits of having them based in a controller, a few lessons have been learned:

- EM templates should use a generic name. During the initial phase of the control system implementation, the first few EM templates

Table 1.2.   Case study two: Time savings for implementation tasks, in hours

Task	Original	Second	Saved
Engineering configuration	178	80	98
Preimplementation testing	32	16	16
Total	210	96	114

were given the name of the first instance. This has created difficulties
when troubleshooting problems and during subsequent implementations when different individuals would perform the work.

- Clarity is important in naming EM instances. A descriptive name
  for the EM instance that leaves no doubt as to its functionality is
  extremely important. Input from Operations personnel is solicited,
  and acceptance of the proposed and agreed to name is crucial. Furthermore, consistency in naming EM instances derived from the
  same EM template has proven to be a time saver.

- Carefully select EM parameters that can be changed and those that
  are fixed. Experience has shown that fewer parameters can be better.
  Fixing parameters that are not directly related to a process, such as
  line purge times, simplifies and minimizes operator interactions.

In addition, many advantages have been achieved by choosing a flexible batch
application that allows EMs to be executed independently of the batch application
manager:

- It has been far easier to deal with problems early on rather than during the production startup of a recipe. In practice, it has been useful
  to have operators manually execute EM logic, catch control logic
  obstacles before recipe implementation, and address them.

- Having EM logic within the controller has allowed the flexibility to
  use EMs to recover from a batch application manager failure.

- Recipe development has been simplified. If the production process
  has been automated to the EM level, then the development of a
  recipe is greatly facilitated. This is the result of the operating instructions being written at the EM level.

Overall, the implementation of EMs independent of a batch application manager has been successful. Basing EMs within controllers has saved time and has
led to more cost-effective subsequent implementations, in addition to the operational advantages of a controller-based application.

An Iterative Refinement Approach toward a Structured Recipe Design

Presented at the WBF European Conference, Barcelona, Spain, November 10–12, 2008, by

Ramayya Kumar
Director
kumar@paperless-gmbh.com
Paperless GMBH, Hirschstrasse 130,
Karlsruhe, 76137 Germany

Abstract

This chapter is a case study within pharmaceutical sites that describes the application of ISA-88 and ISA-95 structures and methodology, in spite of the fact that these sites were using a vendor's product that did not adhere to the standard terminology. An iterative refinement approach was taken while designing the implementation of recipes. Initially a library of process actions was developed for the enterprise. Each process action was comprised of a description, a list of key critical attributes and parameters, and a list of report values. In the first phase, the existing processes and the recipes at the sites were analyzed. Care was taken to focus on the process and ignore equipment related activities. This analysis yielded the overall skeletal structures (general recipes) that essentially reflected the capabilities of the sites. In the next level of refinement, the process operations within these skeletal structures were refined to obtain a network of interconnected process actions. At this stage, certain placeholder process actions were used in the network to represent equipment-related and other procedural activities. Then implementations of the used process actions were performed in the vendor's product. Finally, the recipe was implemented according to the structure that was developed earlier.

This approach significantly reduced the amount of time for deploying and validating thousands of recipes for the various products at the sites. Of course, the time spent designing the initial skeletal structures was longer than implementing a recipe directly, but this approach provided a reusable library and an easier strategy for recipe validation.

Introduction

The pharmaceutical industry has been slow in adopting the introduction of Manufacturing Execution Systems (MES) in comparison to other vertical industries such as semiconductors, automotive, aerospace, food and beverages, oil and gas, and so on. The primary decelerator for this adoption process is concern regarding compliance-related issues. Hence the overall process of changing the validated procedures is a daunting task. In this chapter, we will see how the structured design of recipes alleviates the hardships associated with this change.

What Is MES?

MES can be viewed as the layer sandwiched between Enterprise Resource Planning (ERP) systems and automation systems (Fig. 2.1). The distinguishing features of each of the three layers can be viewed from three different points of view: time, relationship, and "where" and "who" it affects most. In regard to the time perspective, ERP is strategic in nature, MES is operational in nature, and automation systems are instantaneous in nature. ERP plays a major role in the areas of finance and planning, whereas MES controls the shop floor, and automation is centered around equipment. The primary persons and objects involved in the three systems are managers and planners, supervisors and operators, and the sensors and switches, respectively.

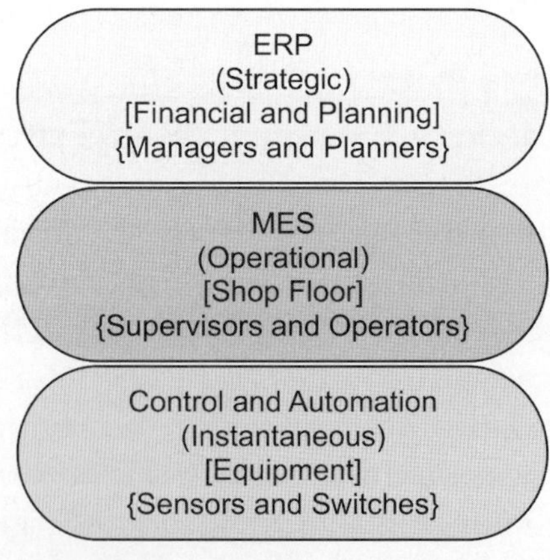

Figure 2.1.   Relationships among ERP, MES, and the automation layer.

Why Use MES?

One of the primary objectives of introducing MES at a site is to reduce variability, thereby improving compliance. An electronic system for capturing the documentation of the manufacturing or packaging processes removes human whims and provides more uniformity. Besides reducing variability, MES also dramatically improves the review of the executed batch records by enforcing the "review by exception" paradigm. Additionally, MES enables users to obtain better visibility of the process under execution. This allows for real-time corrective and preventive actions and also provides the basis for process optimizations.

Who Is Responsible for MES?

Although MES implies a major paradigm shift within the plant and affects practically all the departments, the core MES team should consist of the members shown in Figure 2.2, depending on the scope of the deployment. Subject Matter Experts (SMEs) from the weighing and dispensing and the manufacturing departments are essential if MES is to be rolled out to the manufacturing area. The packaging department SMEs are needed if packaging is a part of the deployment. The quality and the IT departments must always be a part of the core team. Furthermore, management should be a real sponsor of this project and provide utmost support, especially during the initial deployment phase.

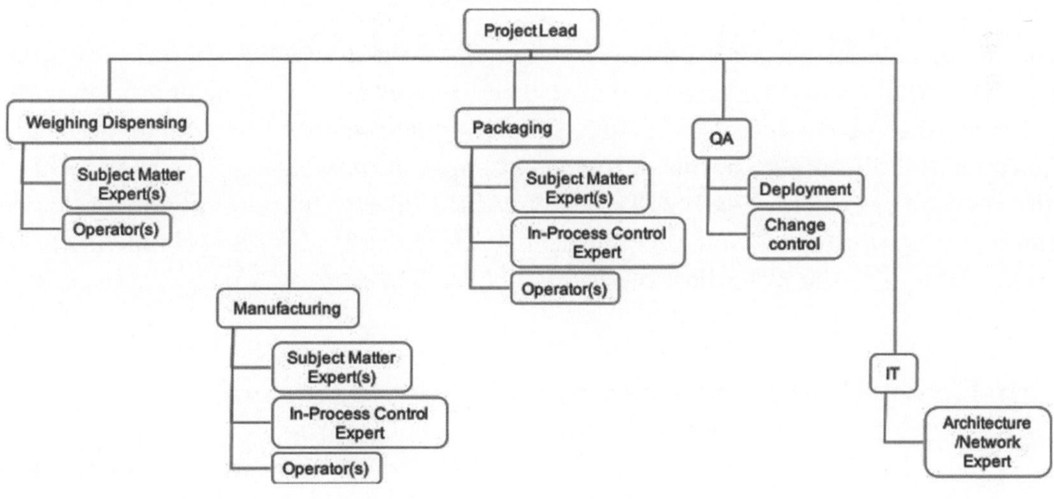

Figure 2.2.   Core MES team members.

The skill set of the members should include good knowledge and experience, good communication and networking skills, and additionally acumen for abstraction and the capability to switch between macroscopic and microscopic views of the process to be modeled.

How Can MES Be Planned?

The questions "where did we come from" and "where do we want to go" allow the site SMEs to examine the current situation and define the scope of the overall project. One of the special considerations from the recipe design perspective is to decide what type of MES you are going to have vis-à-vis the automation layer. The three most popular paradigms are as follows:

- *Paper on glass.* This approach models no connections to the automation layer. In other words, the operators on the plant record the values read from the HMI of the automation.

- *Partial automation.* This approach allows for simple transfers of data between the MES and the automation layers.

- *Complete automation.* This approach sets parameters and then transfers the control over to the automation layer.

How Can MES Be Implemented?

The remaining discussion in this chapter deals with how MES should be implemented with a special emphasis on structured recipe design. Typically the starting point for the recipe design is to use the paper batch record. However there are fundamental differences between the use of paper and an electronic system. These differences are summarized in Table 2.1. Additionally, the conversion of paper batch records into electronic workflow systems provides an ideal opportunity to make changes to the execution processes.

Top-Down Design and Bottom-Up Implementation

Designing the master recipe can be viewed as a three-step process. Initially the conceptual design is performed, which is followed by a blueprint implementation and finally parameterization.

Table 2.1. Paper documentation versus electronic systems	
Paper	*Electronic*
Uncontrolled workflow (i.e., Standard Operating Procedure [SOP]–controlled workflow) is used.	Programmed workflow is used.
Signatures or double signatures are needed for manual entries.	Signatures are needed for the end of work instruction blocks or deviations.
Page numbers, sections, and order-related information is displayed on each page.	All contextual information is available on the current electronic work sheet.
Manual verification of limits and deviation, appropriate signatures, and corrective actions are SOP controlled.	Automatic verification of limits: upon deviations, the control could be electronically programmed or SOP controlled.
Calculations are manual.	Calculations are done automatically.
Material identification (especially checks related to the status) and accounting are performed manually.	Automatic identification of material is performed with real-time check of the status and electronic accounting.
Manual status checks of the equipment are used with corresponding entries in equipment logs.	Automatic checks of the status and log entries are performed.
Parallel execution of steps are used by splitting the sections of the batch record.	Parallel execution is programmed into the model.
Branched execution or alternate steps are SOP controlled.	Branched execution is programmed into the model.
Rework and abnormal scenario handling steps are SOP controlled.	Rework and abnormal scenario handling steps may be partially programmed and partially controlled via SOP.
A review of the complete executed batch records is required.	A prequalification of the electronic batch execution and review by exceptions is required.

Conceptual Design

Creating the recipe blueprints is a conscientious task. This task should be broken down into a conceptual design process followed by implementation. During conceptual design, care should be taken to abstract the overall activities without dependencies on the product and equipment. There should be a focus on the processes corresponding to the chemistry of material transformation. The analysis performed here should disregard all product-related specificities and capture the process-related tasks to be performed. The conceptual design phase can be viewed as a top-down refinement process that starts with the definition of the process operations. The process operations are major processing tasks such as granulation, tabletting, coating, formulation, filling, and so on. These can then be refined to a network of process actions. The process actions represent minor activities to be performed. These could be activities such as mixing, sieving, charging material, and equipment setup. Each of the process actions can be clearly defined with a description and a set of critical and key parameters, attributes, and reporting values. A library of process actions can be built for either the site or the enterprise. This library helps standardize the descriptions, work instructions, reporting parameters, and review.

Blueprint Implementation for Master Recipes

Once the library has been built, the process actions can then be implemented as electronic building blocks that are dependent on the design paradigm (paper on glass, partial automation, or complete automation) that is chosen and on the equipment that is being used. Note that the electronic building blocks can be totally different for each design paradigm. In the paper-on-glass implementation, each instruction requests a manual entry from the operator, whereas partial automation allows the direct push or pull of the values from the automation layer, and complete automation consists of setting only the initial parameters and allowing the batch execution engine to take complete control of the process as defined in a batch recipe. These electronic building blocks can be assembled using the directives from the top-down concept design and parameterized appropriately for creating product-specific recipes. The appropriate assembly of these building blocks defines the workflow for the operators on the shop floor. The building blocks can be interconnected in such a manner that sequencing, parallel execution, branching and looping based on certain conditions, and special triggering of events can be modeled. This interconnected set of electronic building blocks is the recipe blueprint.

Parameterization of the Blueprint for a Specific Product

From a master recipe blueprint, individual master recipes can be derived for each product via parameterization. The electronic building blocks within the blueprint contain only placeholders for each parameter, and these are then specified for each product to create the master recipe for a specific product. For example, a product could have a mixing time and speed of 10 minutes and 160 RPM, respectively; whereas another product's parameters could be 15 minutes and 120 RPM. The recipe blueprints obtained via this design process can be reused via pure parameterization as long as the process is the same for the products. If the process is different, then another blueprint that is a modification of the original blueprint can be used. Thus so-called product families can be built (i.e. those products that can be parameterized from a single recipe blueprint belong to the same product family). This concept will be used later for validation and during the change control process.

Overall Recipe Structure

Every process operation typically consists of the work phases shown in Figure 2.3. In the startup phase, activities corresponding to cleaning and ensuring that the material is present and other procedural tasks are performed. The setup phase includes activities corresponding to cleaning, line clearance, equipment setup, and so on. During the run phase, the typical process activities are executed. The shutdown phase includes the activities for completion of the process operation. Finally, in the reconciliation phase activities associated with the material, yield, and checking the equipment out can be performed.

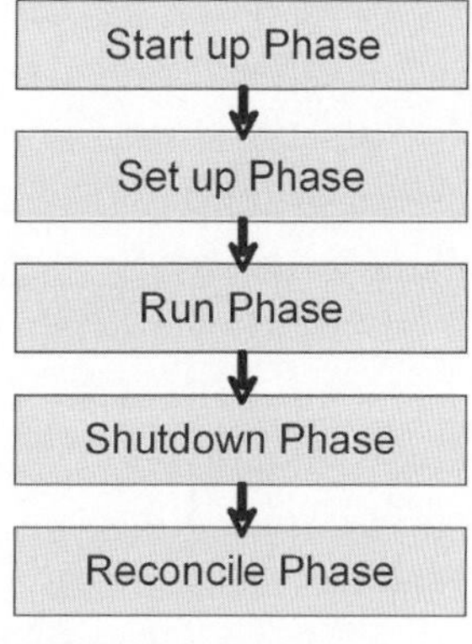

Figure 2.3.   Overall recipe structure.

Figure 2.3 provides a skeletal framework for all the recipes. It is possible that each of these phases may contain either none or one or more electronic building blocks, depending on the process operation that is being modeled. For a better illustration, let us examine the model for a blister packaging process operation (Fig. 2.4).

During the startup phase, procedural activities are performed. The blistering that corresponds to primary packaging and the cartoning that corresponds to secondary packaging mostly run in a kind of pipelined workflow. The overall process operation itself starts with the blistering and is completed with the cartoning. The setup of the blistering line and the cartoning line can be performed at the same time. When the setup of each component of the packaging line is satisfactory, the production run for blistering and cartoning can proceed once again in parallel. Within the production run, normally cyclic in-process control checks are performed on a regular basis. The workflow can also allow the operators to temporarily interrupt the run process and return back to the production run. In such cases, a looped workflow is available for a temporary halt. Finally, the shutdown of the blistering line starts first and can proceed in parallel with the shutdown of the cartoning line. The blister packaging process operation itself ends with the reconciliation of the material used and a yield calculation.

The dry granulation process operation looks very different from the packaging process operation as shown in Figure 2.5. It is essentially a sequential process. The procedural startup activities and the setup activities can be combined into a single electronic building block. This is then followed by a series of building blocks

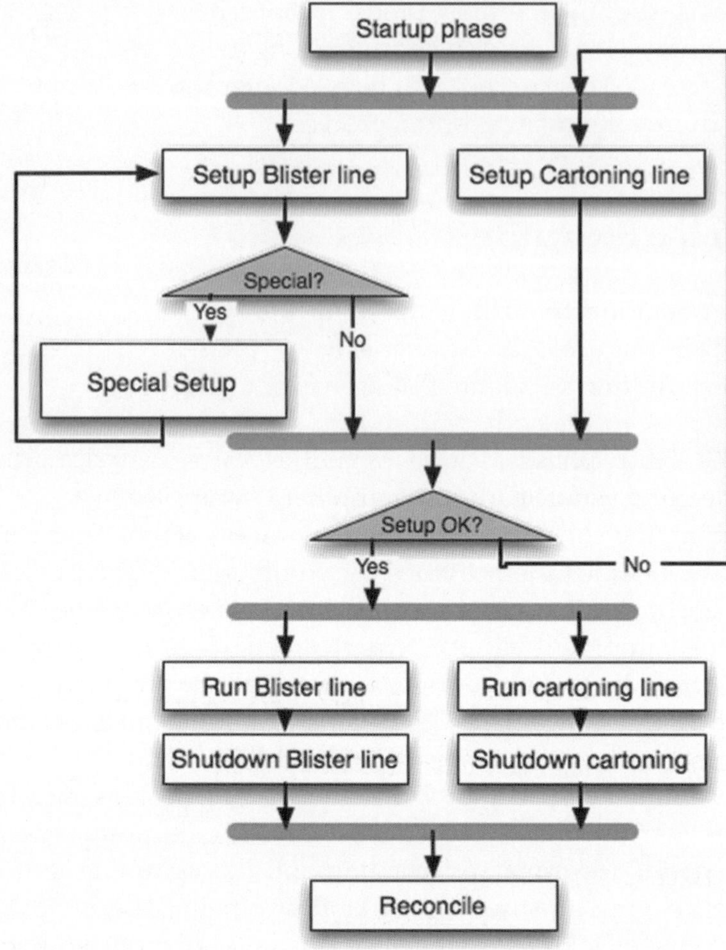

Figure 2.4.   A workflow model for the blister packaging process operation.

that correspond to the run phase. The run phase for dry granulation consists of various material identifications followed by sieving, mixing, and a repetition of these activities with other materials. Finally, the material out building block corresponds to the reconciliation and yield calculations.

Design Validation

The recipe blueprints that have been obtained via the design process described in the previous sections can be validated on a single product to ensure that the workflow and the documentation satisfy the required regulatory, quality, and

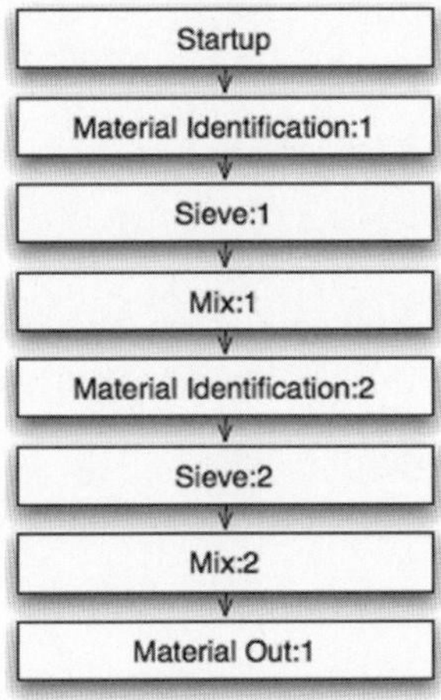

Figure 2.5.   A workflow model for the dry granulation process operation.

operational criteria for the entire product family. This dramatically reduces the design, validation, and approval times. When the recipe for a product belonging to a previously validated blueprint is created, only the parameters need to be verified. When this blueprint has been derived from another validated blueprint, only changes and the accompanying risks need to be validated.

Summary and Conclusion

In this chapter, we have proposed a structured design methodology for recipes within the pharmaceutical industry. It has been shown that a proper abstraction process that focuses on process-related aspects during the initial design phase helps enormously in successfully creating blueprints from which multiple product-specific recipes can be created. Although the initial time for design is much higher than coding individual recipes, the benefits of the time and effort invested in the initial design ensure major savings in the long run.

Applying ISA-88.01 to Small, Simple Processes

Presented at the WBF
European Conference,
Mechelen, Belgium,
November 13–15, 2006, by

Clark L. Case
Requirements Analyst
clcase@ra.rockwell.com
Rockwell Automation, 15458-B N. 28th Avenue,
Phoenix, AZ 85053, USA

Abstract

The benefits of applying ISA-88.01 to large, complicated batch processes have been well established. However, many batch processes are small (involving only a couple of units) and simple, such as blending a few ingredients together. Through an example batch process, this chapter explores some ways that the principles of ISA-88.01 can be applied to such systems. Additionally, it will discuss the benefits that can be derived, such as repeatability, maintainability, and scalability.

Introduction

For large and complicated batch processes that execute multiple large and complicated recipes, following the models defined in the ISA-88.01 standard for implementing the control system is a pretty clear choice. The power of the modularity and "separation of powers" between the equipment and procedures has been demonstrated again and again. However, many batch processes do not fit in this category. In these processes, there is not a lot of equipment involved—perhaps just a couple of units. Additionally, the procedures they perform are rather simple, such as blending together a few ingredients.

For processes such as these, the choice of which design methodology to follow is not as clear. The control system designer may be concerned about the overhead required for a full ISA-88.01 implementation. He may think that to use ISA-88.01, he will have to buy and maintain a lot of additional computer hardware and learn how to use complicated software. For a small and simple process, he might conclude that this is not worthwhile.

What this control system designer might not realize is that ISA-88.01 is not an all-or-nothing decision. Certain benefits can be realized by implementing just part of the model. For example, Control Modules (CMs) and Equipment Modules (EMs) can be implemented without necessarily creating phases or any other elements of the procedural model. This chapter will describe the benefits that can be achieved by implementing various levels of the ISA-88.01 models. It will also explore the drawbacks of not implementing the higher levels.

This chapter is not intended to give a complete description of ISA-88.01 or to be a tutorial in ISA-88.01 design methodologies. There are several good resources for this information, such as *Applying ISA-88: Batch Control from a User's Perspective* by Jim Parshall and Larry Lamb and *ISA-88 Implementation Guide* by Darrin Fleming and Velumani Pillai.

Brief Overview of the Physical and Procedural Models

Two parts of the ISA-88.01 standard are of particular relevance here: the physical model and the procedural model.

The Physical Model (Fig. 3.1) describes how the physical equipment of the process is to be modeled. In part, it defines a Process Cell, which contains all of the equipment required to make a batch of products, including one or more Units. A Unit is a major piece of equipment that performs a major processing step. It may contain one or more EMs. EMs carry out minor processing steps and can contain other EMs and CMs. A CM, which can in turn contain other CMs, does the actual manipulation of the equipment.

The Procedural Model (Fig. 3.2) describes how the procedures used to produce batches are modeled. The top level

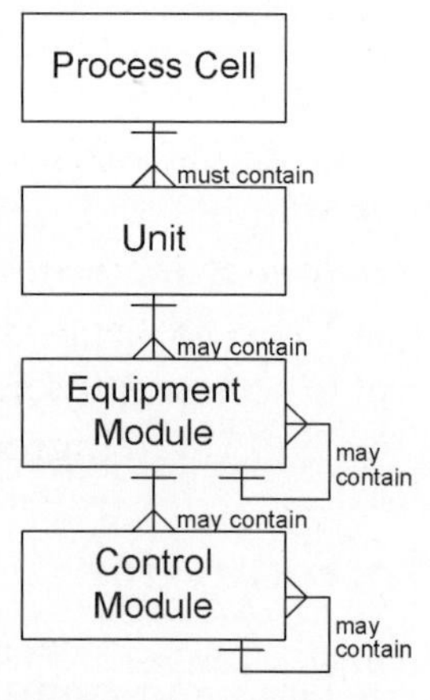

Figure 3.1.
ISA-88.01 Physical
Model (partial).

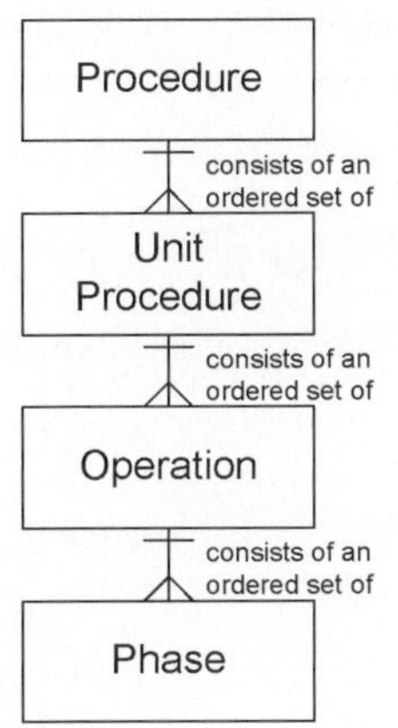

**Figure 3.2.
ISA-88.01
Procedural Model.**

is the Procedure, which is executed in a single process cell and can produce a batch. It consists of an ordered set of Unit Procedures. A Unit Procedure is executed in a single Unit and generally performs a major stage in the production of a batch. It consists of an ordered set of Operations, which in turn consist of ordered sets of Phases. A Phase performs a distinct task in the production of a batch. When implemented in control logic (an Equipment Phase), it can manipulate EMs and CMs.

These are very brief and incomplete summaries of the models. Both of the resources mentioned previously give a more in-depth description and analysis.

Implementing the Models on a Small Process

Most ISA-88.01 design methodologies suggest modeling the equipment and procedures of a process through all the levels of the hierarchies described previously. This holds true for small, simple processes as well. It ensures that the lower-level objects are designed in a way that will allow for easy integration into higher-level objects and procedures, even if they are not used initially.

Implementation of the design in the control system, however, is done from the bottom up: first CMs, then EMs, then equipment phases, and from there either a commercial or custom phase sequencer to implement the procedural model. Additionally, implementation can be stopped at any stage if the requirements of the control system are fulfilled.

The Example Process

An ISA-88.01 design can be implemented using the process shown in Figure 3.3.

This process can be modeled as a single unit. It consists of a jacketed vessel with an agitator, a level transmitter, and a temperature transmitter. The vessel receives three ingredients through individual lines and can use its pump to either recirculate or transfer material out.

Implementing CMs

It is always necessary to implement the CMs. They provide the base level of control for the process. When implementing CMs on the sample process, we might end up with something similar to Figure 3.4.

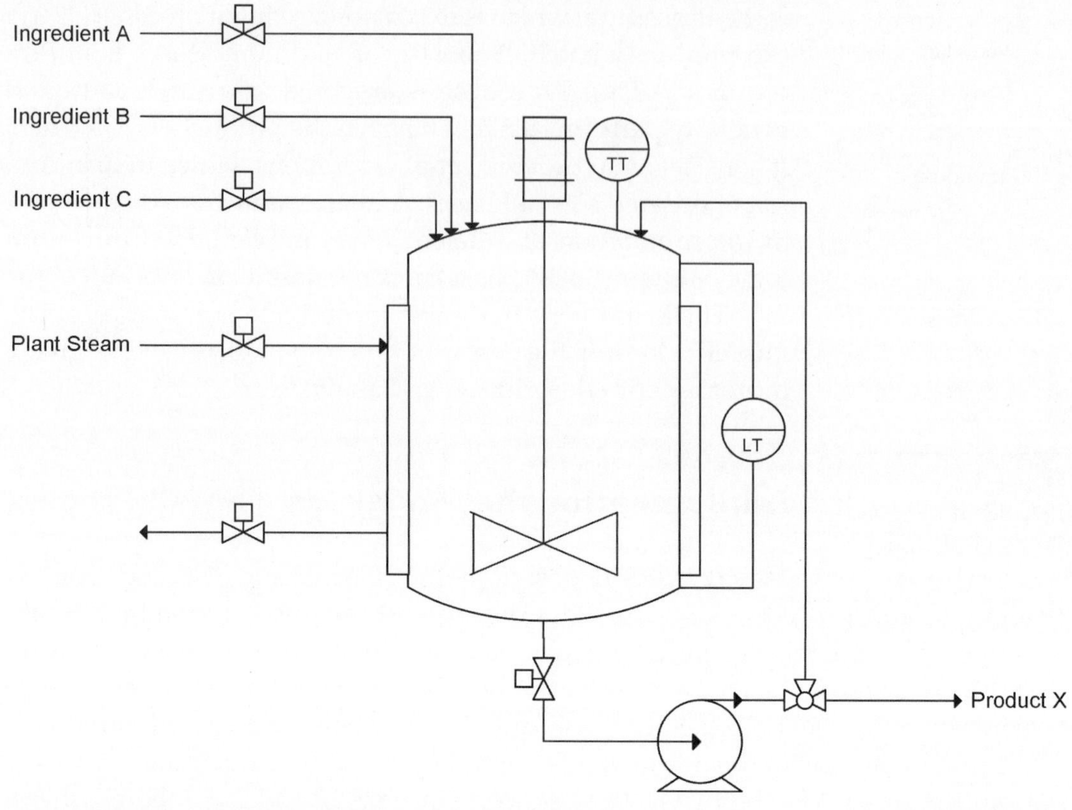

Figure 3.3.   The sample process.

There are CMs for opening and closing each of the valves, for starting and stopping the agitator, and for running the pump. Additionally, there is a CM that uses the steam line valve CM and the temperature transmitter to provide basic temperature control of the tank.

What Has Been Gained

By implementing the CMs, we have achieved several benefits. The most important is process safety. Process interlocks are generally implemented at the CM level. For example, CMs can prevent the agitator from running unless the tank is above a certain level or stop a specific ingredient from being added unless the tank is below a certain temperature.

Another important benefit is failure notification. CMs generally include alarms that would allow, for example, the valve CMs to indicate if a limit switch has not been reached or the temperature CM to indicate that the temperature is too low.

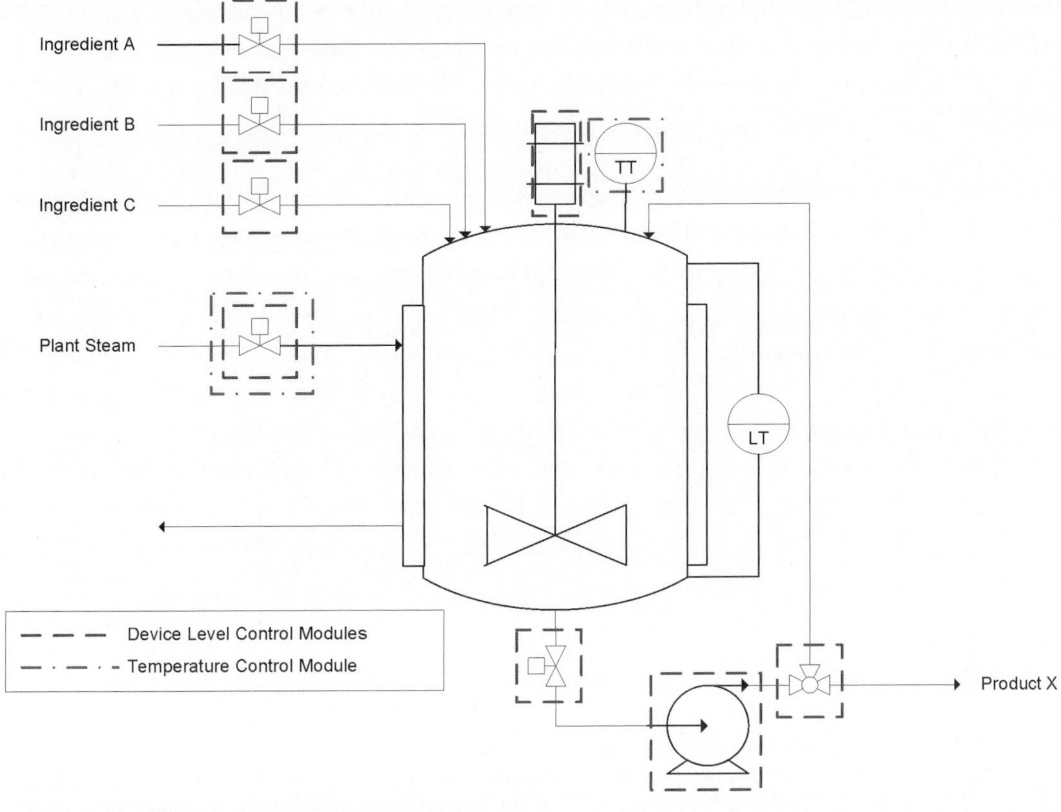

Figure 3.4. Control modules on sample process.

By implementing these CMs, we have given ourselves a foundation for future implementation of higher-level objects. Even if we choose not to do this now, when and if it becomes necessary, the CMs will be available for use with EMs and equipment phases.

Why We Might Stop Here

It could be entirely possible for us to stop our ISA-88.01 implementation here. We could create faceplates for the CMs, and operators would set the temperature, position the valves, and run the pump and agitator as necessary.

There are several possible reasons for doing this. It might be a temporary step while the EMs, equipment phases, and recipes are being developed. In fact, running the process from the CMs would probably provide a lot of information for the development of these components. Producing a batch may take a long time, and manipulation of this process is very infrequent. In this case, it may be deemed that creating the higher-level objects is not worthwhile. This process may also be such that it is never run the same way twice, and device-level flexibility is always required. If this is truly the case, the higher-level objects would not be terribly useful.

Drawbacks of Stopping Here

Of course there are some important drawbacks to stopping the ISA-88.01 implementation at this point. If this is a typical batch process running multiple batches per day, the process is going to require intensive operator interaction. The operator will constantly be opening and closing valves and monitoring conditions. This will necessitate having very highly skilled and trained operators. Even then, chances are that batches will not be produced in a very repeatable manner.

Additionally, it will require more effort to generate batch data. A typical data historian can be used to collect data on a continuous basis, but converting these continuous data into information summarized on a per-batch basis—such as "73.2 gallons of Ingredient C were used in batch 531"—will be difficult.

Implementing EMs

The next level of implementation for our ISA-88.01-based control system design is the EMs. These give us, where needed, coordination of CMs to allow them to work as a group. For the example process, we would probably have two EMs (Fig. 3.5).

One EM coordinates the three-ingredient addition valve CMs and watches the level transmitter to see how much has been added. The second EM coordinates the vessel's outlet valve CM, the pump CM, and the recirculation three-way valve CM. It also watches the level transmitter to indicate when it is finished pumping out the tank.

What Has Been Gained

Because the EM-level logic takes care of the coordination of the CMs, we have significantly reduced the amount of interaction the operator must have with the processes. For example, he can simply command the ingredient addition EM to add 30 gallons of Ingredient B. This replaces his previous process, which included making sure no ingredient addition valves were open, making sure the outlet valve was closed, noting the level of the tank, opening the Ingredient B valve, waiting for the level to raise 30 gallons worth, and then closing the Ingredient B valve. The processes of recirculating for a certain period of time and pumping the tank until it is empty are similarly simplified.

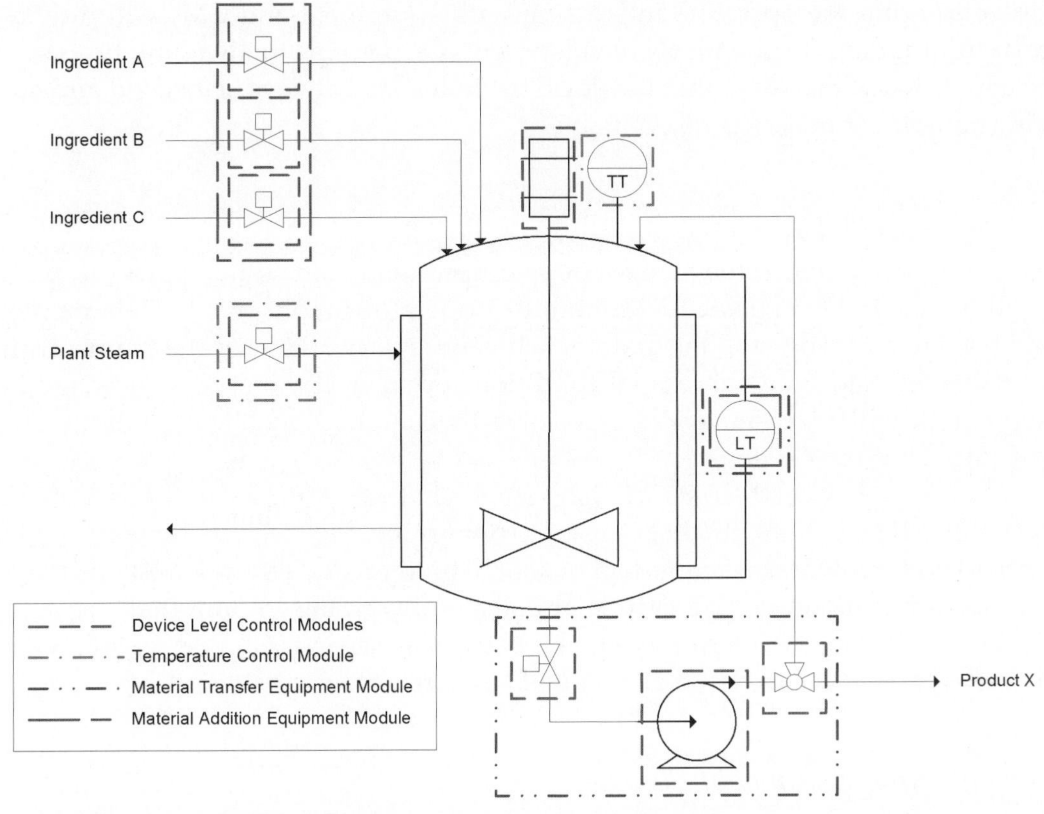

Figure 3.5.   EMs on sample process.

By reducing the required amount of operator interaction, we have increased repeatability. We also may have reduced the amount of operator skill and training needed to run the process. Also, as we did with the implementation of CMs, by putting in EMs we have a basis for future equipment phases and a deployment of the procedural model.

Why We Might Stop Here

At this point, we could create a few more faceplates for the EMs and start running the processes. The operators would command the ingredient-addition EM to add ingredients, use the temperature CM and agitator CM to heat and mix, respectively, and use the transfer-out EM to recirculate and pump out the tank.

The reasons for stopping here are similar to the reasons for stopping at the CM level. The equipment phases may be under development. Again, if that is the case, observing the operators interacting with equipment and CMs will provide great information for the phase development. Or perhaps the way the process is executed changes so often that EM-level flexibility will always be needed and creating equipment phases is not valuable.

Drawbacks of Stopping Here

While we have reduced the amount of operator interaction with the process, there is still going to be a substantial amount of operator involvement. The operator will have to start and stop the agitator when necessary and keep the temperature set correctly. And if any abnormal conditions occur or if the process needs to be stopped, he will have to know how to correctly command the EMs to put the process into the proper state.

Although having the EMs will help somewhat, generating summarized batch data will still be a chore. It will be easier to determine, for example, how much of a particular ingredient was requested, but some thought and effort will be needed to associate that amount with a particular batch and with the amount that was actually added. A way to determine when a batch was started, when it ended, when abnormal events occurred, and so on will also need to be considered.

Implementing Equipment Phases

Equipment phases are phases implemented in sequential control logic. These are typically the highest-level objects that reside in the controller and are usually the interface points for commercial batch sequencers. They manipulate EMs and CMs to carry out a step in the production of a batch. The set of equipment phases for our sample process might resemble Table 3.1.

Table 3.1.	Sample process equipment phases		
Equipment phase name	*Objects used*	*Input parameters*	*Output parameters*
Add_ingredient	Ingredient addition EM	Ingredient, Amount	Amount added
Agitate	Agitator CM	Speed, Time	Actual time
Heat	Temperature CM	Temperature, Hold time	Maximum temperature
Recirculate	Transfer out EM	Time	Actual time
Transfer_out	Transfer out EM		Actual time

What Has Been Gained

Because the equipment phase logic manipulates the equipment and CMs as needed, the operator does not. This further reduces the amount of operator interaction in the process, and as we have seen, this means that less operator skill and training is needed.

Generally, equipment phases implement the ISA-88.01 recommended set of commands (e.g., Start, Stop, Abort, and Hold) and the recommended phase "state machine." The state machine defines the possible states of a phase, such as Running, Holding, and Held. It also defines the ways a phase can get from one state to another. This gives our operators a small and finite set of ways to interact with a phase and gives the phase developer a way to handle those interactions.

For example, the equipment phase developer will put logic in the phase to handle the Hold command. The Hold command is used to put the phase into a state where it is no longer running but can later be restarted. While the phase is processing that logic, it is in the Holding state, and when it is done it is in the Held state. The developer will also put logic in the phase to allow it to handle the Restart command. This command starts the phase up again from the Held state. While this logic is being executed, the phase is in the Restarting state, and after it is completed, the phase is back in the Running state.

Equipment phases also generally include input and output parameters. The input parameters are used to specify what the phase should do, and the output parameters allow the phase to report what actually happened. The input and output parameters, combined with the predefined set of commands, give the operator a concise way of manipulating the processes. To add 75 gallons of Ingredient A, for example, the operator will set the Amount input parameter of the Add_ingredient phase to 75, set the Ingredient input parameter to Ingredient A, and give the phase the start command. If the operator needs to completely stop the phase prematurely, he can give the phase the stop command. If not and the phase runs to completion successfully, the phase will write the amount of Ingredient A actually added to the Amount_added output parameter.

Why We Might Stop Here

After the equipment phases have been created, we could create a display with a table of our phases that allows the operator to specify their input parameters, output parameters, and a set of buttons for the operator to command the phases. To create a batch, he might run the Add_ingredient phase a couple times, then run the Heat and Agitate phases, and then run the Recirculate phase followed by the Transfer_out phase.

As with CMs and EMs, our stop here might be only temporary. Executing the phases in this manner could be a good method of figuring out the best way to write our recipes. Perhaps the order in which phases are executed varies so widely that writing static recipes does not make sense. It is also possible that the process being executed varies so little and is so easy to run at the phase level that the effort of writing a custom sequencer or the costs of buying a commercial sequencer cannot be justified.

Drawbacks of Stopping Here

If we stop here, the operator is essentially acting as our phase sequencer. So, while the phases themselves will be executed in a very repeatable way, the sequencing of the phases may not be repeatable. The possibility also exists of the operator entering parameter values incorrectly, which could result in batches that must be scrapped.

The operator also must monitor the status of the phases and know how to react to abnormal conditions at the phase level. If a phase goes into the Held state because of the failure of a piece of equipment, any other phases running at the same time should probably be put into Held as well.

The speed of recipe execution could also be a problem. When one phase is complete, an operator may need to start another, and no matter how fast he is at starting the next phase, he probably will not be as fast as a computer doing the same job.

With the phases' input and output parameters, the job of collecting batch data has become somewhat less painful. However it will still require effort to come up with a scheme that can associate the data, as well as information on when the various phases started and stopped, when batches started and stopped, and so on, with a particular batch.

Creating a Custom Formula Manager and Phase Sequencer

If we want to get our operator out of the phase-sequencing business, one thing to do would be to write a custom formula manager and phase sequencer. The way this can be implemented varies widely, but essentially the job of an application of this type is to load a formula (or set of parameter values related to a particular product) into the input parameters of the equipment phases and execute the phases in a particular order.

Advantages of a Custom Formula Manager and Phase Sequencer

Once our formula manager and phase sequencer are deployed, they will take care of the phase interactions. This means that, in general, the operator's involvement is reduced to working with our application to start batches. This of course gives us better batch-level repeatability and a higher level of product quality.

Also, since it is a custom application, our formula manager and phase sequencer can be made to fit our needs completely. It can be deployed in the controller with the user interface in the Human Machine Interface (HMI) or as a stand-alone personal computer application. It can be made to handle just one phase sequence or multiple phase sequences. If needed, it can be made to handle all our batch data collection.

Drawbacks of a Custom Formula Manager and Phase Sequencer

Writing our formula manager and phase sequencer will, of course, take time. If the requirements rise above the fairly simple (i.e., if it is to execute more than a few different phase sequences, collect extensive batch data, or interact with a higher-level system), the costs of development should be carefully balanced against the cost of a commercial batch sequencer.

Custom phase sequencers often do not scale well and require significant code changes when new processing equipment is added. This is particularly true if equipment is shared between parallel units, such as a pump that can be used to transfer material out of two vessels. Commercial batch sequencers have built in ways of "arbitrating" this equipment.

Using a Commercial Batch Sequencer

Another way to automatically sequence our phases is to use a commercial batch sequencing package. These packages will allow us to define the ISA-88.01 procedural model items—Procedures, Unit Procedures, and Operations—that describe the way we produce our batches. Operators can then command the batch sequencer to execute these recipes. The batch sequencer then executes phases in the manner defined in the recipe. The recipe for our process could probably be defined as an Operation-level recipe. As a sequential function chart, that Operation might resemble Figure 3.6.

Advantages of a Commercial Batch Sequencer

As with the custom phase sequencer, the commercial batch sequencer allows the operator to primarily concern himself with managing batches. When things are running well, the lower-level objects will not require much of his attention. When things are not running well, he will still be able to take control of the equipment phases, EMs, and even CMs in order to make any necessary changes.

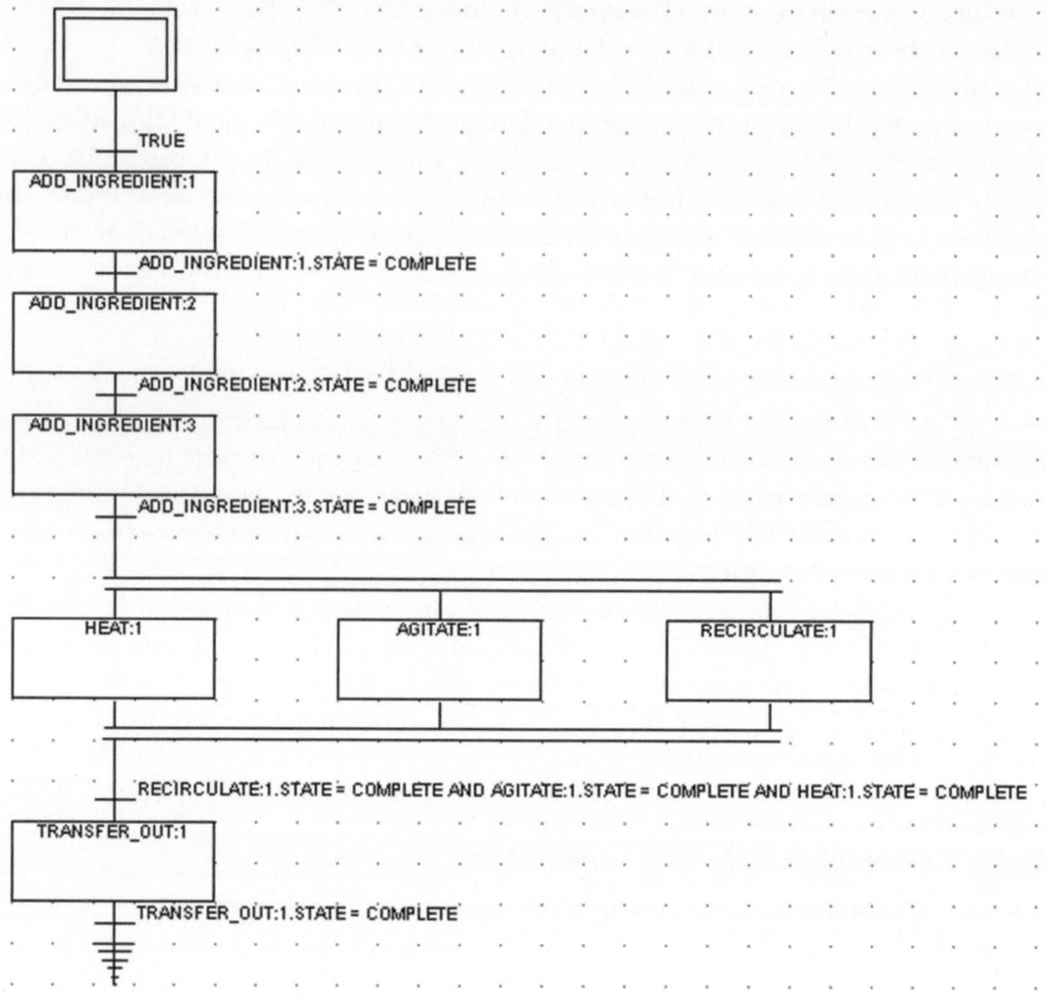

Figure 3.6. Recipe for the sample process.

If we use a commercial batch sequencer, there will be no custom code to debug and maintain. This will allow us to focus on tweaking our recipes to make them run their best. The commercial batch packages allow for the creation of recipes in a very flexible manner that would be difficult to replicate with a custom phase sequencer.

All of the commercial batch sequencers take care of collecting detailed batch-level data. For a given batch, we will be able to know which phases ran, when they started and stopped, what the input and output parameter values were, and so on. This is particularly important when working in a regulated industry.

As mentioned previously, it is easier to adapt to changes in the process equipment when using a commercial sequencer compared to a custom phase sequencer.

New units and phases can be quickly added to their physical model of the process, and the new phases can be easily incorporated into existing recipes. If needed, existing low-level recipes such as the Operation for our sample process can be incorporated into Unit Procedures and Procedures. No custom code changes are required.

Drawbacks of a Commercial Batch Sequencer

There are several costs involved in using a commercial batch sequencer. First, there is the cost of the software itself. Second, there is the cost of the hardware needed to run the software and the cost to maintain that hardware. Depending on the IT infrastructure, these costs can be significant.

There is also a training cost involved. Engineers must learn how to define the physical model, create recipes, and interface the sequencer with the equipment phases. Operators must learn how to interact with the sequencer to create and manage batches. These costs must be balanced against the costs of the other solutions presented and against the benefits that can be derived.

Conclusion

We have looked at the benefits that can be derived from implementing CMs, EMs, equipment phases, and phase sequencing for small and simple processes. The pros and cons of stopping at points along the way have also been explored. In conclusion, the physical and procedural models of ISA-88.01 can and should be applied to even the smallest batch process.

ISA-88.01: A Business Model or an Engineering Implementation

Presented at the WBF
North American Conference,
Baltimore, MD,
April 30–May 3, 2007, by

Lynn W. Craig
President
l@lcraig.com
MAA Inc., 10 Strafford Circle Road,
Medford, NJ 08055, USA

Abstract

We have long thought of ISA-88.01 as an engineering standard dedicated to the technical aspects of automation of batch processes. However, it can also be viewed very properly as a business model for the "shop floor" or Physical Manufacturing Operations (PMO) level of a manufacturing enterprise. This business view is important for many reasons, including broader understanding of the role, resources, and requirements of the level within the business where people and equipment actually manipulate and convert raw materials into products or intermediates. Considering that more of a company's resources are invested in and used by the operations at this level in comparison to the rest of the enterprise in total, understanding the workings and structure of this part of the business should be an obvious necessity. Yet too often the operations at this level are overlooked or misinterpreted as part of an engineering domain complicated by obscure automation technologies and unrelated to "real" business operations. This chapter will attempt to show the usefulness of the ISA-88.01 models and concepts as the core of a business model that describes the operational structure, requirements, and interactions that take place at the plant floor level.

Introduction

The original goal of the ISA-88.01 standard was to provide an engineering tool that would clearly define a common model and methodology for the control or automation of most types of batch manufacturing processes. Such an undertaking required common terminology to allow the exchange of information about the process and the way it is to be controlled. Common terminology, in turn, required models of the batch manufacturing environment that would be abstract and flexible enough to be applicable to essentially any batch manufacturing process. The ISA-88.01 standard was intended to serve as an engineering tool, and it serves that function very well. In the process of describing a batch manufacturing environment, an obvious but largely overlooked by-product of the effort emerged: from a slightly different viewpoint, it is also a good business model for an important level of a manufacturing enterprise. After all, control and automation exist to make a manufacturing process function "the way it should" (as defined later on in this chapter). Following this thought, a tool that allows for a description of the way the manufacturing process should run also allows an operational view of the process itself.

The Need for Abstraction

Any description of a manufacturing process in a way that can be applied across differing equipment and a wide range of functionality must, by definition, be an abstract representation. ISA-88.01 does that pretty well. Having good abstract models of the physical assets—the equipment—that make up a batch process is necessary but insufficient: an abstract view of function is equally important. Without an understanding of function, it is difficult to visualize the equipment doing anything. ISA-88.01 accomplishes that by allowing abstract definition of the combination of equipment capability and the desired functionality. Equipment functionality is, of course, provided by control, automation, and people. The abstraction provided by ISA-88.01 helps immensely if the reader tries to understand the distribution of duties between automation and people. It quickly becomes obvious that functionality can be automated or can be manual and that, within limits, it does not matter whether the functionality is created manually or through automation. The functionality resulting from automation or manual control is essentially interchangeable. Since control of a batch process requires (and will probably continue for many years to require) a mixture of manual control and automation, there will be some of both in any implementation. Rather than trying to sort out the automated portions in order to define them in detail, the standard's abstract models help to define all required functionality first. This approach can aid manufacturers in decisions about the optimum degree of automation that can better be dealt with as implementation options.

The need for generalization and abstraction is obvious but difficult to attain. Clearly it is only through abstraction that models can be developed that can be applied to a broad range of actual batch processes. When ISA-88.01 was written, control as we knew it primarily focused on basic or regulatory control. This kind of control was well founded in theory, broadly used in industry, and widely available commercially. What was not totally clear at the time was the extent of manual activities performed by human operators. In manually controlled processes, people were doing a pretty good (if not necessarily precise or overly swift) job of making the processes run. Although there was no concerted effort by the standards writers at the time to rigorously examine precisely what these people were actually doing, a fair amount of time could probably have been saved if that route had been pursued. Instead, an appreciation of their activities was more of an intellectual exercise for the writers, and to some degree this appreciation came indirectly from certain writers who had spent years soaking in the culture and practices of manually controlled batch manufacturing plants. Looking at the activities of operators in a manual plant is still a good way to understand the procedural and coordination requirements described in the standard. It is also a lesson in optimization. Manual control of chemical and process manufacturing practices is not new and is definitely not yet obsolete. Such manufacturing processes have been in existence for many years, and the culture and organizational structures that control these processes have evolved and matured over the years. This evolution has defined, refined, and optimized a workable structure with multiple levels of responsibility and has gone a long way toward defining the duties of the people who fill the various levels in that structure. A production manager has one set of responsibilities, a foreman has another, and an operator has yet another. The culture has established a hierarchy and a well-defined suite of responsibilities for each level. It is that optimized hierarchy of function and responsibility that ISA-88.01 has captured with a fair amount of accuracy.

Operators and Process Control

Human operators typically work on one or more major pieces of equipment (Units) and know how to make their part of the process operate. A closer examination of what operators actually do reveals that their duties require them to know how to perform predefined process-oriented tasks or procedures in a sequence appropriate for the product being made. Although operators may manipulate equipment directly, in modern plants they are more likely to perform procedures by manipulating basic controls that set equipment or process values to a desired condition or state. Operators typically set the states in some predetermined order to carry out a fairly well-defined task. The individual tasks are usually done the same way regardless of the product being manufactured. Thus operators manipulate basic controls to carry out reasonably well-defined procedures in an order necessary to make a given product. The order of the procedures they perform in a totally manually controlled plant are typically based on a "batch card" or some other written sequence of procedures that is much like a recipe that one might use in a kitchen. Also, operators informally coordinate activities among themselves to avoid contention for resources and keep things running relatively smoothly. In addition to that, there is usually a foreman present who is responsible for the entire process, its people, its effectiveness, and its safety. The foreman is also dedicated to keeping the work (or procedures) of the various operators coordinated.

The obvious lesson from all this is that basic control—or to use the categorical term, Process Control—was defined well before ISA-88.01 and only needed to be represented in a way that meshed with the standard. Individual small procedures could then be defined to command Process Control to carry out predefined process-oriented tasks. Considering the need for product-specific sequencing of tasks, another level of procedures could be defined to determine the desired order in which the process-oriented tasks performed by operators or automation should be performed for specific products. In addition, if a process is to be automated, a method should be developed that takes into account the need for coordination.

The Standard Emerges

From insights gathered by people over many years of experience and with only about five additional years of meetings needed to get consensus, the ISA-88.01 standard finally emerged. It was based largely on the now familiar physical and control activity models. It has definitions and terminology based on these and several other models such as the procedure, process, and recipe-equipment linkage models. The standard also recognizes the relationship between the real-time functions carried out in the physical processing equipment (Process Cell) and the management-level activities that were later defined in more detail as Manufacturing Operations Management in the ISA-95.03 standard. All of this was done with a degree of abstraction that not only allows the standard to be applied broadly but also allows processing operations to be visualized without concern for whether they would be implemented manually or with sophisticated automation systems.

The standard defines a structure that is abstract and still relatively easy to envision with minimal study or training. The structure described in the standard can be implemented, but the standard itself is definitely not an implementation specification. It is constructed in a way that allows the structure to be implemented, but it leaves open the means and details of implementation. The ease with which the structure can be envisioned is of major importance and can provide a useful management tool. If a combination of manual and automated control is to be enacted in reality, there is a vital need for the people who define and are responsible for manual tasks to be able to communicate accurately with the engineers who must create the automated capability for the process. In reality, implementation is—or at least should be—an operations management issue. The people who are responsible for operating the plant need to understand what is to be implemented and ensure that the procedures and other controls will match these responsibilities. Once the users have defined what is needed, they can implement the manual operation and training of people, and the automated part of the implementation becomes an engineering issue. It is important at both the operations and engineering levels that the definition of the function comes first and clearly defines the functionality each group must bring to the party. It is only after the development of a clearly understood definition of the way the plant should run that implementation can be done well. It is also very important for the implementation of both manual and automated control to resemble this definition. Otherwise the interactions between the two will be lost, and further improvement will be hampered.

Defining Process Control

The whole purpose of control is to make a process operate like it should. Properly defining the most desirable way for a process to run is, at best, difficult. To do it properly requires a team approach with control engineers, operations people, and others. Obtaining a consensus from these groups can take time and slow down engineering progress. It is all too easy for engineers to avoid this difficulty and define automation more or less unilaterally. This is a natural progression from the typical control engineering culture. No one else usually has a clue about how or why to create a cascade control loop or to apply model reference control. The control engineer has, historically, simply had to do it. Usually no one objected because they did not have the technical know-how to challenge the engineering approach. It is tempting to follow this same approach with the automation of procedures and other batch control issues. After all, someone must define the way the process should operate and usually the pressure is on the engineering staff to show progress. In addition, the operations staff often does not want to get tangled in the engineering details. This approach often seems to work but leaves the people who are responsible for operating the plant on the outside, often not even looking in. This is a mistake that can easily happen if communication between the various stakeholders is difficult.

There is no good reason why people who know how a process should run cannot work with engineers to define exactly how it should run with automation. Everyone involved understands equipment, and everyone understands function. ISA-88.01 deals with both, and if everyone sticks to this standard, interaction and communication are not only possible but actually easy. Often there is the temptation to use a control system representation as a basis for conversation. This often fails because such representations are typically provided by a control system company and are intended for engineering customers. While these representations are beneficial to the engineering community, they can confuse communications with people who have less of an engineering background. Both the engineers and the operations people need to understand the ISA-88.01 models but should definitely avoid getting bogged down in the technically rich representations of control systems that companies provide for their engineering customers. Both need to use the ISA-88.01 terminology as a common language and avoid the technically derived categorizations provided for automation implementers. Only then can a control strategy actually emerge and optimum ways to operate the plant be devised. This strategy and agreement defines what control must accomplish.

Using ISA-88.01 to Define Process Control

When used as described previously, ISA-88.01 models and terminology take on a different role. Viewed in this way, they are a reference model for the physical processing level in a manufacturing business. They provide a way to visualize and evaluate changes in operating procedures at that level. They are an operations tool that is very useful for understanding and analyzing that level. But what is that level? It does not really have an agreed-upon name. It is often called the "shop floor," which conjures up an image of a twelve-story building with the CEO at the top and a nasty workshop, looking sort of like an anvil factory, being run in a subbasement. Another term that is used is "plant floor," which veers away from the anvil factory image but still does not even hint at the complexity and financial impact of that level of a manufacturing business. We fortunately have a name for an enterprise and a manufacturing site. We can finally talk about the functionality and level in a company that ISA-95 has defined as Manufacturing Operations Management. However at the level where people and machines touch and modify raw materials to create something else or carry out procedures to make equipment ready to do something else or carry out tests to verify changes in raw materials, we are stuck with terms like "shop floor" or "level one and two," neither of which describe much of anything. Let's face it—until we gained the ability to automate procedures in a fairly elegant way, Manufacturing Operations Management and the confusing welter of stuff that function managed was lumped together in most people's minds and treated as "something" that somehow made things. ISA-95 has helped define and describe the level that manages manufacturing operations and clarified its role. Now we need a descriptive name for the "stuff" this level is managing (Fig. 4.1).

ISA-88.01 defines it but does not name it in any definitive way. Let us be bold and invent a name that bears some relationship to what is going on at that

Top (CEO) level

use money to make more money

Enterprise level

spend money to provide services and activities to help make money, such as
Accounting, Marketing, and Sales

Research and Engineering

spend money to invent things

Manufacturing Operations Management

figure out how best to make what sales has sold/will sell and spend money to
understand it and do it better and less expensively

Physical Manufacturing Operations

use money, procedures, people and equipment to convert raw materials into
products or intermediates

Basic Control

use invested money to set equipment and process condition states - largely
an engineering domain

Figure 4.1.   Informal levels in a manufacturing business.

level—at least until we have finished with this chapter or someone has come up with a better name. The name that comes to mind is "Physical Manufacturing Operations" (PMO). It even looks OK as an acronym, as long as we do not mix it up with the Prime Minister's Office, Palermo International Airport, or Project Management Offices. It seems compatible with Manufacturing Operations Management but still indicates the difference. After all, what goes on at that level is no longer management. It is all about physically touching raw materials and equipment and carrying out preordained tasks. It is the focus of what seems to usually be called "operations" or "manufacturing operations" in most companies.

Alright—we have had fun, been arrogant, and invented our own name. Why is that important? Why do we need to identify a previously ambiguous level in an enterprise? Well, while it is closely tied to Manufacturing Operations Management, it is a level that can operate on its own for significant periods of time based only on information that was provided at an earlier time—perhaps like instructions sent on Friday afternoon before management goes home for the weekend. It has its own culture that has been refined and optimized over many years. More of a company's resources are invested in and used by the operations at the PMO level, when compared to the rest of most manufacturing enterprises in total. It is a level that directly impacts not only cost but also quality, capacity, reliability, yield, safety, and so on. It is a pretty important level (Fig. 4.2).

Other than the definitions in ISA-88.01, the boundaries of the PMO level are not generally well understood; the extent of this level has never been well described, especially in the processing industries. Until automation became an issue, people in these industries did not need to know about this level. With total reliance on the

Figure 4.2.   PMO relationship to Manufacturing Operations Management and basic equipment control.

manual control of procedures, there is no good reason why PMO needs distinction from other manufacturing-oriented functions. Dealing with people can be casual and need not be all that precisely organized. In a totally manual case, precision does not matter much. People will generally figure out what is meant and what they need to know with no harm done.

Automation has changed this. Computers do not do well without precise definitions. That is one of the important things about the ISA-88.01 models and terminology. They are precise enough to define both automation and manual functionality without cluttering up definitions with technical details that hinder understanding. It is important that we recognize the need for specific definitions and keep our models and terms straightforward and easy to understand. The over-all ISA-88 family of standards contains four approved parts and one more part that is under construction. More could very well come along.

ISA-88.01 is an excellent technical and engineering tool. However, it is different from the other three parts. Most of the parts must be formally defined with engineering precision and are focused on engineering and information notations and technicalities, unlike ISA-88.01. By its very nature, ISA-88.01 is conceptually oriented and defines principles, models, terminology, and functions that can be understood without a detailed knowledge of control engineering or information technology. It is the portal though which engineers and operations people can communicate, collaborate, and deal with PMO together. Keeping the relative clarity of ISA-88.01 for all stakeholders is vital and necessary in spite of the ever-present engineering temptation to make any control-related standard more detailed and dedicated to precise control engineering and information-technology requirements.

Bottom Line

ISA-88.01 is not an engineering implementation. ISA-88.01 is an engineering tool, and it is also a useful business model for the PMO level in almost any process manufacturing enterprise. The fact that it discusses both business and operational requirements makes it a workable reference model of a fairly complex level in the manufacturing business. Viewed in this way, ISA-88.01 is a management tool and a basis for defining automation from an operations viewpoint. It provides a basis for defining the optimum level of automation in modern manufacturing. It serves as a basis for communication between people at different levels in a company and provides the tools for evaluating performances across different sites and different management approaches. It can also be a useful management tool to help identify and remove communication barriers.

The ISA-88.01 Area Model: More Than Just a Pretty Face

Presented at the WBF North American Conference, Philadelphia, PA, March 24–26, 2008, by

Doug Bourgeois, PE
President
doug.bourgeois@cs-automation.com
Complete Systems Automation,
1955 Cliff Valley Way, Suite 220,
Atlanta, GA 30329, USA

Scott Sommer, PE, CAP
Automation Technology Manager
scott.sommer@jacobs.com
Jacobs Engineering Group,
2 Ash Street, Suite 3000,
Conshohocken, PA 19428, USA

Editor's Note

ISA-88.01 does not have an Area Model. It has an Area in the Physical Model with a few words to describe it. The author's use of "ISA-88.01 area model" appears to refer to the application of the models, terminology, definitions, activities, principles, and practices contained in ISA-88.01 to a project as a design model. Please read "area model" in that way.

Abstract

Hundreds of major manufacturing projects begin every year, many of which require some level of batch automation and control. ISA-88 principles are employed on almost all of these batch automation projects, due to its acceptance as the prominent framework for defining batch automation. In the authors' experience, following an ISA-88 approach provides benefits from project conception through startup and validation.

It has also been the authors' experience that most batch automation projects advance too far before the definition of the process control model is started. Often project decisions made prior to the process control model definition place limits on the ability of the systems integrators and batch automation professionals to provide the most efficient, value-added batch automation configurations. The full benefits of ISA-88.01 are not realized if project managers do not include the definition of the model as a required document during the Basis Of Design (BOD) phase of a project.

This chapter will present actual project data demonstrating that the definition of the ISA-88.01 area model in the BOD phase of a project is vital to the following:

- Planning and scheduling of automation activities throughout the project
- Quantifying the number and scope of documents required
- Identifying the scope of Systems Integration services required
- Developing the proper dependencies between Automation, Instrumentation and Control (I&C), and Construction
- Providing the most complete and realistic cost estimate
- Understanding the commissioning, qualification, and validation efforts required
- Providing a basis on which to validate the system architecture, system boundaries, panel boundaries, and control schemes

Throughout this chapter, the benefits and value of defining the ISA-88.01 area model during the BOD phase of a project will be identified.

Introduction

The objective of this chapter is to explain the importance of developing the ISA-88.01 area model during the Preliminary Design Phase of a project, commonly referred to as the BOD phase. The traditionalist view—or more appropriately, the typical standard view—of industry today is that the required information is not available at this stage of a project and subsequently there is no value added by defining the area model during the BOD phase. Most system owners wait until the System Integrator comes on board to define the area model, but this chapter argues that this is too late in the process.

In this chapter we will (1) detail how our objective applies to different phases of a project, (2) explain the rationale of why we believe we need to break the traditional way of thinking, and (3) reveal the benefits of defining the ISA-88.01 area model during the Preliminary Design or BOD phase of a project.

Project Phase Deliverables

Project deliverables will vary based on industry, type of project, and specific client. However in order to understand the proper place in the project timeline for the ISA-88.01 area model to be developed, it is necessary to generally understand the development of project phase deliverables.

Traditionally, during the Conceptual Phase, the Business Drivers (or Business Case) are developed to justify the expenditure of funds for the project. Once this phase of the project is funded, a Process Flow Diagram (or Process Block Diagram) is developed, along with corresponding Architectural Floor Plans to support the process and required support functions. Once this is accomplished, permitting strategies, construction strategies, and utilities requirements are addressed.

Ideally, during the Conceptual Phase, an Automation Vision document is prepared to identify the role of instrumentation, automated systems, and information flow for the proposed process and architecture. The end product of this phase of the project—other than the documentation of the previously listed activities—is a project milestone schedule and a factored cost estimate. While it is too early in the process to develop an ISA-88.01 area model during the Conceptual Phase, it is not too early to begin to define the methodologies that will be used to analyze the process systems and to identify the role that the ISA-88.01 (and ISA-95) standards and concepts will play in the development of the automation for the project.

The Preliminary or BOD phase of the project is where the groundwork is set for all the disciplines involved in the project. Specification templates are developed; typical drawings are produced for everything from control panels, HVAC ductwork, and pipe racks to underground plumbing, incoming utilities, and architectural room finishes. This is the phase in which Equipment Specifications are developed, long lead items are specified and ordered, and Issue for Design Piping and Instrumentation Drawings (P&IDs) are developed. From a construction standpoint, permitting packages are submitted, detailed cost estimates are developed (usually falling within ~10% of the overall costs), and Level 2 schedules are developed during this phase.

From the automation engineer's standpoint, User Requirement specifications, System Architecture diagrams, and Instrumentation specifications are produced during the BOD phase. In addition, support is given to process engineering in the development of preliminary Sequence Of Operations (SOO). It is during this project phase that development of the ISA-88.01 area model should take place. We believe that the ISA-88.01 area model is just as important as frozen architectural floor plans or Issue for Design P&IDs.

The importance of developing the ISA-88.01 area model during the BOD phase becomes initially apparent when the deliverables and major milestones required during Detailed Design are identified. During the Detailed Design phase, the Process Sequence of Operations is completed and plant personnel begin to draft standard operating procedures. P&IDs are issued for construction, and equipment specifications are issued for bid. Coordinated spatial designs confirm that architectural, mechanical, and piping systems do not clash in pipe racks, under slab ducts, within wall systems, in mechanical spaces, and so on. The construction manager is hired and produces a Level 3 schedule. Construction permits are issued, bid packages are issued for the various trades, and subcontractors are brought on board.

At this point, the automation engineer is worried about producing Functional Specifications, issuing the System Architectures for construction, and issuing the bid package for System Integration. In short, the design is complete and bid packages are "on the street" in anticipation of constructing the plant according to the project timeline. Unfortunately, in the authors' experience, this is usually the point in the project where attention is paid to the ISA-88.01 area model. Waiting until this point in the process to focus on the ISA-88.01 area model is a leading cause of software rework, schedule "busts," and cost overruns during the execution phase of the project.

Role of the ISA-88.01 Area Model in the Project Timeline

As described previously, developing the ISA-88.01 area model during the Detailed Design phase usually means that P&IDs are nearly complete, instruments have been specified and are issued for bid, the System Architecture is nearing completion, and the System Integrator is coming on board. However, isn't it the System Integrator's role and responsibility to define the ISA-88.01 area model? So shouldn't the project wait for the System Integrator to be "on board" before developing this model?

The authors' answer is a resounding "No!" Their experiences over the past ten to fifteen years have shown that developing the ISA-88.01 area model at this late stage of a project has the following negative effects:

- Systems Integrators have to "work around" the system architecture to meet recipe requirements (e.g., data flows not supported, too many peer-to-peer connections, improper amount of hardware specified, capability of hardware components lacking, too many custom interfaces required, etc.).

- Improper process segmentation and undefined or poorly defined system boundaries are disconnected from automation boundaries (e.g., Input/Output [I/O] segmentation, controller scope, etc.).

- Excess numbers of or not enough Equipment Modules (EMs) and Control Modules (CMs) are required based on process and instrument design (e.g., no thought given to reusability, consistency, class-based design, etc.).

- Workflow considerations are not addressed (e.g., improper placement and number of operator stations, lack of planning compensated for by added procedural logic that complicates operator interactions, etc.).

To counteract these issues, the authors believe the following: to achieve the most efficient, value-added batch automation implementation, the ISA-88.01 area model must be developed during Preliminary Engineering (BOD) phase.

While the previously stated information may convince an experienced automation engineer to champion the development of the ISA-88.01 area model at an earlier stage in the project cycle, the challenge is convincing project managers about the benefits of spending the time and funds during the BOD phase to define the ISA-88.01 area model. In order to do this, the following sections will identify specific benefits that early ISA-88.01 area model development will bring to a project.

Using the ISA-88.01 Area Model to Develop the Optimal Batch Control Strategy

This section will describe how the ISA-88.01 model can be used to facilitate the optimal batch control strategy. This will be done by examining the role of the ISA-88.01 model in its various parts: the Physical Model (Units, CMs, and EMs) and Procedure Model (Phases, Operations, and Recipes).

The "optimal" control strategy can be defined as the batch control strategy, which incorporates process requirements, user requirements, and project execution requirements (e.g., project management, construction, and scheduling). This should be implemented in such as way as to not require functional specification, hardware specification, or software specification changes in the Detailed Design phase of the project. In other words, the optimal control strategy is developed by coordinating the automation requirements with all other disciplines. This requires the development of the ISA-88.01 area model during the BOD phase of the project.

Physical Model

The Physical Model should be developed in conjunction with the P&ID development effort. This coordination will ensure that the proper EM and CM classes are developed for modules of similar functions. Representations of control loops and device control can be standardized to facilitate the definition of EM and CM classes. Once these classes are defined, the appropriate control system inputs and outputs to support these CM and EM classes can also be defined.

With the User Requirements as a guide, the proper control strategies can be represented in the P&IDs. In conjunction with constructability reviews and commissioning or validation requirements, system boundaries, control system segmentation, and the logical division of process descriptions can be determined. System boundaries and process control system segmentation tasks are the keys to determining the number and placement of control panels, the number of controllers and the functions required in each, and the placement of interface panels and Human Machine Interface (HMI) stations.

These activities, tasks, and definitions will be refined, tested, and adjusted during the BOD phase. Performing these tasks in the BOD phase produces accurate cost estimates and baseline schedules before continuing with detailed design activities.

Procedural Model

As with the Physical Model, development of the Procedural Model during the BOD phase can provide tangible benefits. The procedural model includes recipes, unit procedures, procedures, operations, and phases. These elements of the ISA-88.01 area model are often ignored by automation professionals during the BOD phase because they generally require the automation engineer to participate in process discussions, to interview operators and manufacturing personnel, and to work with the facility's disciplines, such as architecture, piping, and process mechanical engineering.

Development of the procedural model requires an understanding of material and personnel workflows, operator tasks and work zones, manufacturing support functions (laboratory, material kitting, and maintenance, to name only a few), and manufacturing data flow requirements. Only when these non-software-related and non-control-related topics are fully addressed and understood does the automation engineer have the information required to adequately define the elements of the Procedural Model. From these inquiries, the automation engineer can fully participate in the development of Process Descriptions, verify sequences of operations, locate operator terminals in the appropriate locations, understand the optimal data entry methods (e.g., barcode, full HMI terminal, wireless tablet PC, handheld terminal), and identify unique functions and procedures that must be addressed.

Not to be lost in the BOD phase tasks is the development of skeletal recipes that mimic intended plant operations. These allow the automation engineer to validate the results of the process segmentation effort, determine which operations and modules should be shared, determine the extent of interactions between units and EMs, and determine the extents of phase, EM, and CM classes.

The main benefit of completing these exercises during the BOD phase is the avoidance of penalties later on in the project life cycle. All the aforementioned tasks will be performed during a project, either explicitly (as the authors suggest) or implicitly (as a reaction to software modules that just do not work, modules that cannot be validated, or modules that fall short of user requirements in function or interaction). In these cases, the old adage of "pay to do it right the first time or pay twice to fix it in the field" applies.

Field fixes undoubtedly lead to kluged software, unstable application codes, project delays, and added costs. The Project Manager who does not value the time spent up front to develop the required area model definitions in the BOD phase may initially be heralded for keeping the engineering costs down on the project. However, the additional costs and negative schedule impacts incurred during the execution phase makes the construction manager look bad and is what, in the authors' view, is responsible for much of the bad press automation engineers get when things do not go well once automation is on the project's critical path.

Benefits of Early Development and Deployment of the ISA-88.01 Area Model

This section details the benefits of the early development and deployment of the ISA-88.01 area model. These benefits, in no particular order, are based on the experience of the authors who have completed or participated in many ISA-88-guided projects from conception to commissioning. In some cases, we had to initiate some of these benefits in order to realize them. We are certain there are also additional benefits that have yet to be recognized.

Project Unity

In a typical large-scale project, many different disciplines from many different companies with many different management philosophies are involved. For the purpose of this argument, let's restrict these unique disciplines as such:

- Engineering
- Automation
- Validation
- I&C
- System Integration
- Construction
- Operations
- Skid Vendors

These various entities often deploy their own project management and lead personnel to organize, manage, and track their respective project deliverables. One can easily see that with the many players involved, different ways of doing things, and various methods of documenting design, the ISA-88.01 area model can be used to unite these differences and form a universal language and structure that is understood and implemented by all players.

Improved Cost Estimation

Generally a cost estimate (within 10%) is required in the BOD phase of a project. For this benefit analysis, let's limit our investigation to automation deliverables. In order to accurately estimate automation costs, one needs to have a detailed analysis of the total number of system components ranging from I/O count, EM and CM templates, Unit classes and respective Units, Phase classes, Recipes (e.g., Operations, Unit Procedures, Procedures), software and licenses, system hardware (e.g. control panels, controllers, HMIs, historian servers, batch servers, OPC [Object Linking and Embedding for Process Control] servers), and so on. The System Architecture drawings, I/O list, and SOO, which are available in the BOD phase, provide detailed information that is beneficial for cost estimating. The following is a list of additional requirements:

- Quantify the number of ISA-88.01 Physical Model elements
- Quantify the number of ISA-88.01 Procedural Model elements

- Quantify the number of supporting documents required for the two preceding model elements, which can be a major cost item

- Align process boundaries and automation boundaries (because misalignment is a validation nightmare and one of the causes of budget overrun in industry today)

- Detail Process Cell to Process Cell transfers of material (and data)

- Detail unit to unit transfers of material (and data)

- Detail peer-to-peer communications (e.g. skid-mounted Programmable Logic Controllers [PLCs] to Distributed Control Systems [DCSs])

- Take into account validation strategies or methodologies for process boundary definitions

The early release of the ISA-88.01 area model will detail this information and allow for a more realistic and complete cost estimate.

Improved Documentation

Let's observe the simple example of a bid package and associated Request For Proposal (RFP) document issued to an automation vendor at the beginning of the Detailed Design phase. Assume that the System Owner requires a Functional Specification and a Detailed Design Specification for each Unit class. This is very typical in the industry at the time of this writing. If the RFP does not detail the ISA-88.01 Physical and Procedural Models, then how can the total number of workflow documents and code-development man hours accurately be accounted for in the vendor proposal? Just how does the automation vendor determine how many Unit class Functional and Detailed Design Specifications they are supposed to bid? Many vendors make these decisions based on educated guesses. However, as is common in today's competitive market, vendors are reluctant to supply a price for something that is not accounted for in the RFP. The reason for this oversight is that they may be penalized for the additional cost, which could then possibly give an unfair advantage to their competitors. It is the ethical responsibility of the System Owner or their representatives to perform an apple-to-apple comparison between different vendor proposals, but sometimes the ever-mighty lower cost is too alluring.

Some might ask, Can an accurate and complete proposal from a System Integrator be realized consistently if the ISA-88 area model is not defined prior to the bid packages being issued? Most System Integrators are well versed in ISA-88.01 and can read between the lines with the P&IDs, System Architecture drawings, and

SOOs to develop their proposals. However, it has been the authors' experience that if the System Owner does not detail the Physical and Procedural Models in the RFP, then scope creep is inevitable during Detailed Design. It is imperative that the ISA-88.01 area model be developed before bid packages are issued to ensure all the major components of the design are accounted for in vendor proposals, whether hardware or software related. Although this means System Owners must spend more money in the front end of the project, they will ultimately save money and time in the back end by minimizing scope creep.

The early release of the ISA-88.01 model can also assist in the development of the SOOs. SOOs can sometimes be difficult to translate, as their language and terminology are not consistent, and they do not follow ISA-88.01 guidelines. Typically what is defined as an Operation in the SOO is not defined as an Operation in the ISA-88.01 Procedural Model. Educating the authors of the SOO on the Physical and Procedural Models can only help them create a more complete, structured, and meaningful document. The physical structure of the documents should be modeled after the ISA-88.01 area model, in which case a Phase equals a Phase, and an Operation equals an Operation. This is not rocket science, but rather it just needs to be taught to all participants from the Preliminary Design phase through the Detailed Design phase so that an apple in the BOD phase is an apple in the Detailed Design phase.

More Efficient Detailed Design Phase

The development of the SOO occurs in conjunction with the development of the P&IDs. The SOO is used to develop the Functional Specifications in the Detailed Design phase. The SOO becomes a historical document once the information has been transposed into the Functional Specifications. If the ISA-88.01 area model is developed by the System Integrator during the Detailed Design phase, it must be a predecessor in the project schedule to Functional Specification development, which is a predecessor to the Detailed Design Specification development. Failure to define an accurate ISA-88.01 area model at this point in the process can cause substantial project delays. It has been the experience of the authors that many of the required disciplines are not consulted (or only briefly consulted) during the ISA-88.01 area model development in today's typical project execution. It is not uncommon for project managers to assume that the System Integrators will get it right the first time, as they typically are the ones with the highest level of expertise. Imagine the case in which the System Owner's Validation department is not consulted during the ISA-88.01 area model development because the project manager fails to orchestrate this or sees no value in it. There exists a risk that the Automation boundaries as defined by the System Integrator do not match the Process boundaries. In simpler terminology, the Validation department refuses or has difficulties validating the system based on the architecture, as it not aligned with their past methodologies. In most cases, Validation wins and this causes substantial rework during the Detailed Design phase.

Requires Validation Input Earlier

The implementation of the ISA-88.01 area model should involve the system owner's Validation department. The automation boundaries have a direct impact on their scheduled activities and cost estimating. If the ISA-88.01 area model is developed in the BOD phase, then automation and validation alignment is realized before the Detailed Design phase is initiated. This will minimize the risk of incompatibility between process boundaries and automation boundaries.

Well-designed Automation Code

It should go without saying that complete and accurate documentation and a well-thought-out ISA-88.01 Physical and Procedural Model translates to better code and better traceability. If the basic controls of the automation process are defined early, then these controls can be referenced during P&ID and SOO development. In simple terms, use the ISA-88.01 language as early as possible, as it will make for an easy transition from SOO to Functional Specifications to Detailed Design Specifications and ultimately to code development. Failure to define adequate or enough EM or CM classes leads to a great deal of code customization and generally complicates the validation process. System Owners need to take the responsibility to ensure that adequate process control templates are available and not rely on System Integrators to create these templates on the fly.

Critical Tasks for Maximizing the Benefits of the ISA-88.01 Area Model

Let's have some fun and play the "Thou Shall" game. Basically, this section briefly highlights some of the critical tasks that automation engineers must consider if they expect to maximize the benefits of the ISA-88.01 area model. Remember, this is the foundation for the entire Automation system so it pays to get it right. Although one could easily argue that it pays to get to it wrong also.

Thou shall

1. define Process Areas;

2. organize Piping and Instrumentation Drawings (P&IDs) and work-flow documentation, schedules, budgets, and so on;

3. align all aspects with validation.

Thou shall

1. define Process Cells within Areas,

2. define material transfer and data flow between Process Cells,

3. define material transfer and data flow between vendor-supplied skids (PLCs) and Process Cells.

Thou shall

1. define Unit classes and Unit instances within Process Cells;

2. organize workflow documentation likewise (SOO, Functional Specifications, Detailed Design Specifications, etc.);

3. consult with validation about the methodology of validating Unit classes, Units, and Skids.

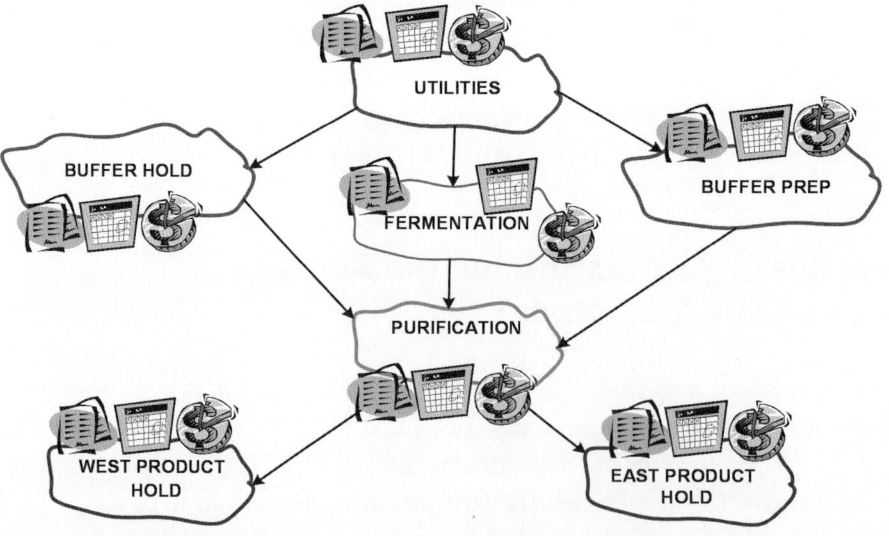

Figure 5.1.   Define Process Areas.

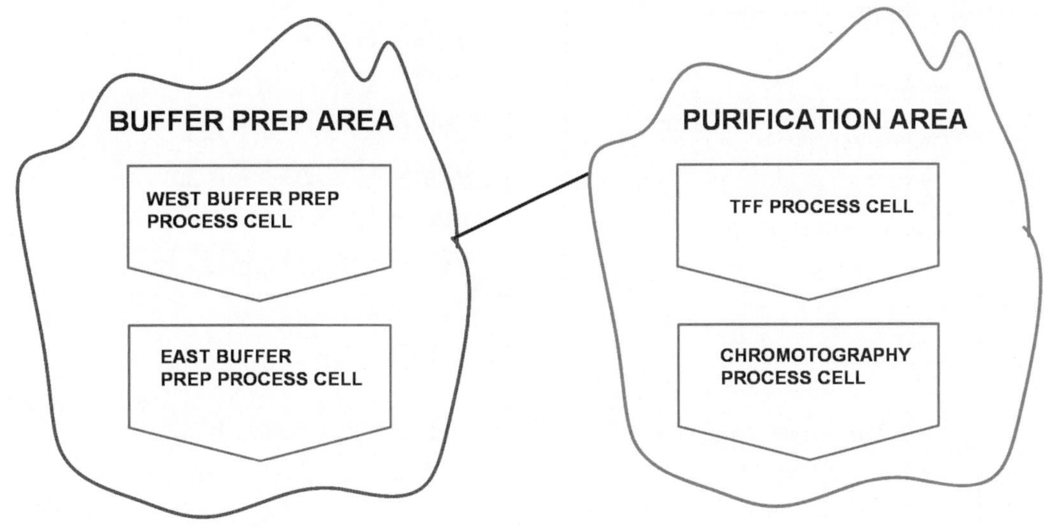

Figure 5.2.   Define Process Cells.

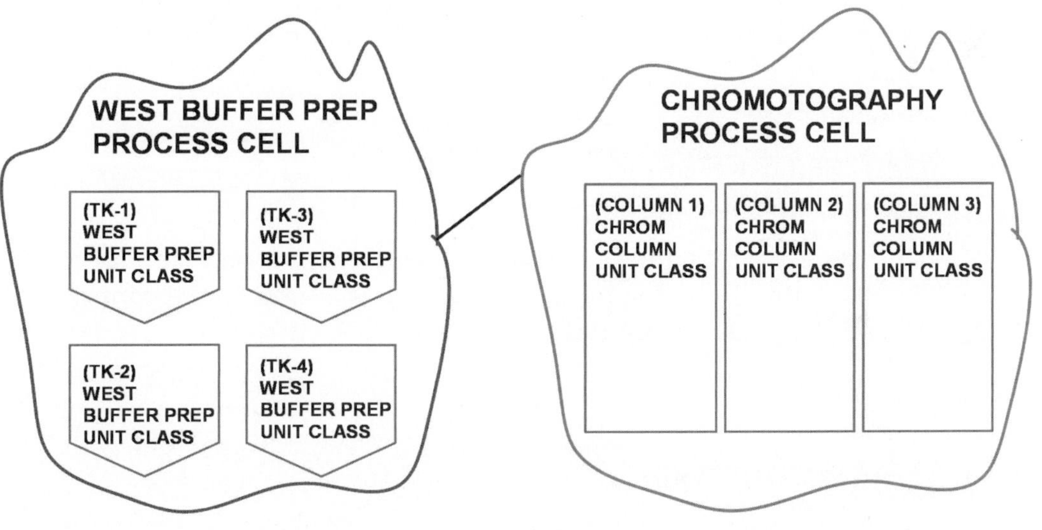

Figure 5.3.   Define Unit classes.

Thou shall

1. define Phase classes and their association with Parent Unit classes;

2. define Phases as Phases in SOO and not as Operations;

3. understand the difference between a Phase and an Operation.

Figure 5.4.   Define Phase classes.

Thou shall

1. define EMs and understand their importance at all levels of the project implementation, including SOO development in the BOD phase;

2. define all EMs, even for trivial control schemes;

3. understand that the formation of EM classes generally means less customization of code later;

4. understand that the more you can segment your process and use EMs to control these segments, the easier it is to implement control and validate your configuration.

Thou shall

1. identify shared resources;

2. analyze potential bottlenecks.

Conclusion

The ISA-88.01 area model is a very powerful method for designing process control systems. Lately, in various industry types, it is becoming popular not only for batch processes but also for continuous processes. The primary reason for its wide acceptance in industry is that it has a proven track record, and quite frankly,

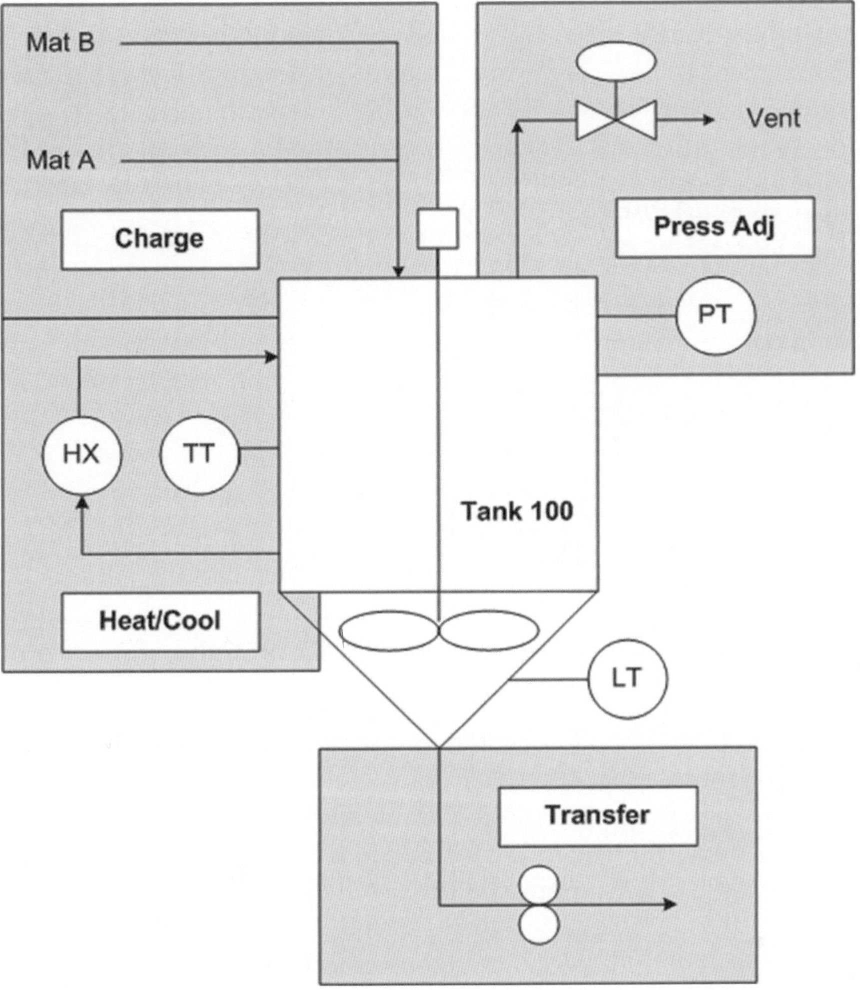

Figure 5.5.   Define EMs.

it works. The WBF reports that various industries have derived the following benefits from using ISA-88.01:

- 20% reduction in the cost of automated systems
- 30% reduction in time to implement and change systems
- Reduction in time to develop product definitions (from days to hours)
- A reduction in the standard deviation of batch cycle times
- 33% reduction in the number of required operators

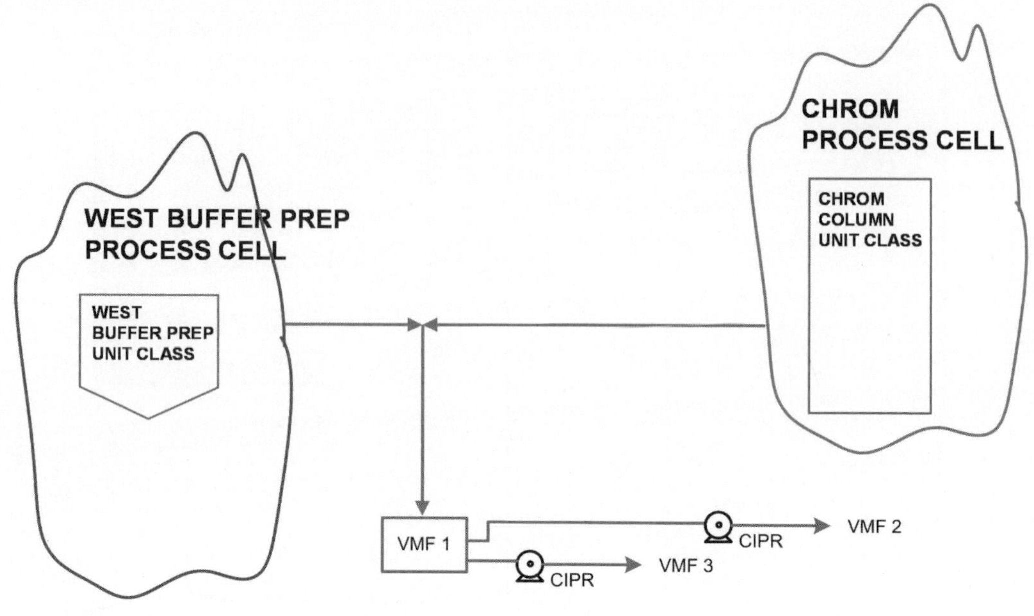

Figure 5.6.   Identify Shared Resources.

The authors of this chapter argue that it is time to promote the ISA-88.01 area model to an even more important role in project execution. It has proven its success in the Detailed Design phase, and now it is time to prove its worth in the Preliminary phase of project execution. The following benefits can be realized by promoting and deploying the ISA-88.01 area model up the project execution ladder:

- Standardization of project management activities across different departments and companies

- Complete and realistic cost estimation

- A common hierarchy, structure, and content of workflow documentation

- Development of the proper dependencies between Automation, Validation, I&C, and Construction

- Improvement in the quality of information presented to the Detailed Design phase developers

- A more efficient Detailed Design phase

- Improved code

- Improved system architecture

Flexible Recipes, ISA-88.01, and Control

Presented at the WBF
European Conference,
Barcelona, Spain,
November 10–12, 2008, by

Francis Lovering
fl@controldraw.co.uk
ControlDraw Ltd
Portsmouth, United Kingdom

Abstract

Some plants can support multiple products being simultaneously manufactured in an array of equipment supported by highly flexible routing. These are the flexible plants that this chapter addresses. ISA-88.01 provides a good framework for describing the equipment and the recipes of these plants, but in practice the control software can become highly involved and inflexible, especially around the transfers.

Introduction

This chapter is about the sort of flexible plants that support multiple recipes and have multiple flow paths connecting the equipment. This chapter specifically focuses on the control of these flow paths. First, it provides examples of such transfers, and second, discusses how these have been handled in real applications. It explains how inflexible some real applications really are from the point of view of the control engineers who design the systems. Some plants manage to achieve a good deal of flexibility from the point of view of the recipe writers but at a huge expense in terms of the effort required to program the automation.

Flexible Plants

Flexible plants can have many Units and Equipment Modules (EMs), many of which are functionally the same and can be described by classes. There can be many Material Transfers between them (e.g., by valves including multi-port valves, by manually connected hose stations or pipe bends, and even by temporary connections).

As an example (Figs. 6.1, 6.2, and 6.3, which are just a portion of a much larger plant), the inter-unit connections are made by a combination of valves and pipe bend connections. There are dozens of transfer routes and even more Clean-In-Place (CIP) routes (Fig. 6.1). In the hose station depicted in Figure 6.2, the routing is highly flexible, and the manifold multi-port valves depicted in Figure 6.3 allow for multiple simultaneous routes. The following example of a hose station (Fig. 6.2) shows its great flexibility. If the hose station is replaced by multi-port valves, then the possibilities for automation are increased (Fig. 6.3).

Describing the Requirements

The plant requirements can be described fairly easily using ISA-88.01 models and terms to describe each Unit type. So if there are several similar Units, it is only necessary to describe one using ISA-88.01 terminology to define a class. This even applies to material transfers when referring to the source and destination in generic terms. But it gets difficult when it becomes necessary to describe the control of each possible transfer route.

What Does ISA-88.01 Say?

ISA-88.01 does not help with transfers, as there is very little in either the original or even the latest update about such connections. It does say the following:

- "The control of what equipment to allocate to the different batches, and when transfers can take place may require control at the Process Cell level."

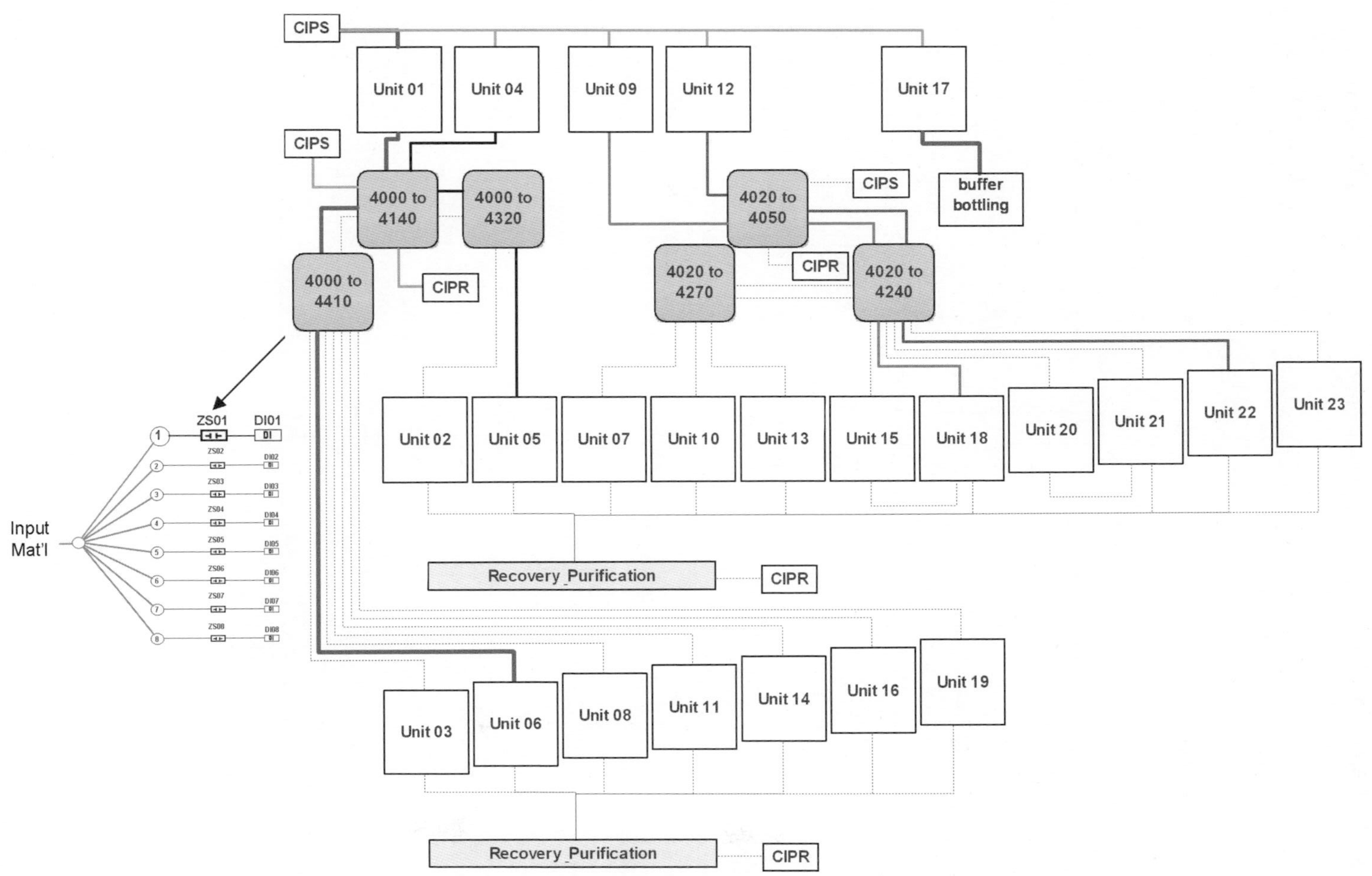

Figure 6.1.   Portion of a flexible plant.

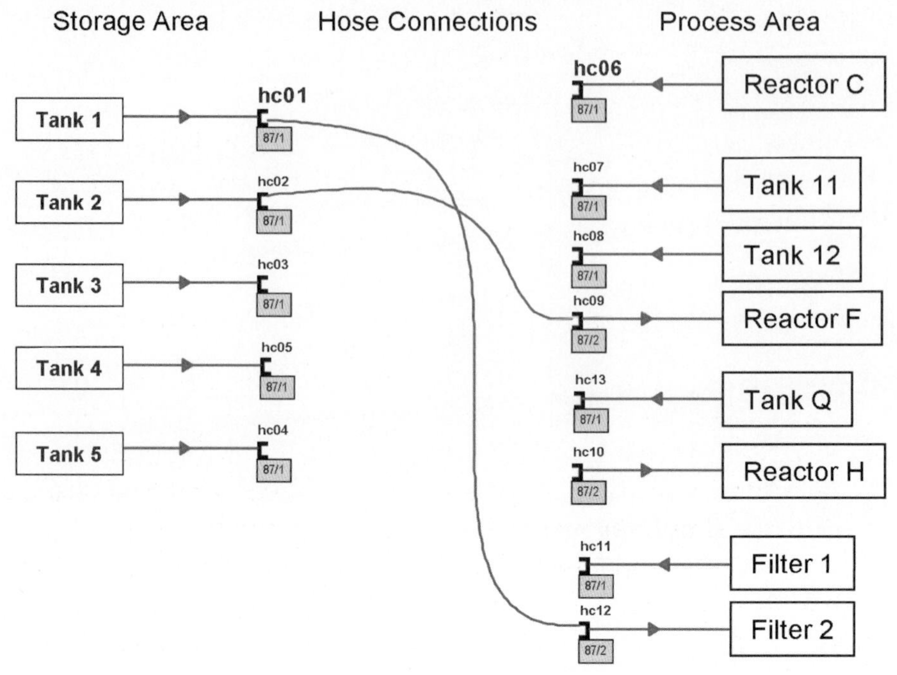

Figure 6.2.   Example of hose station routing.

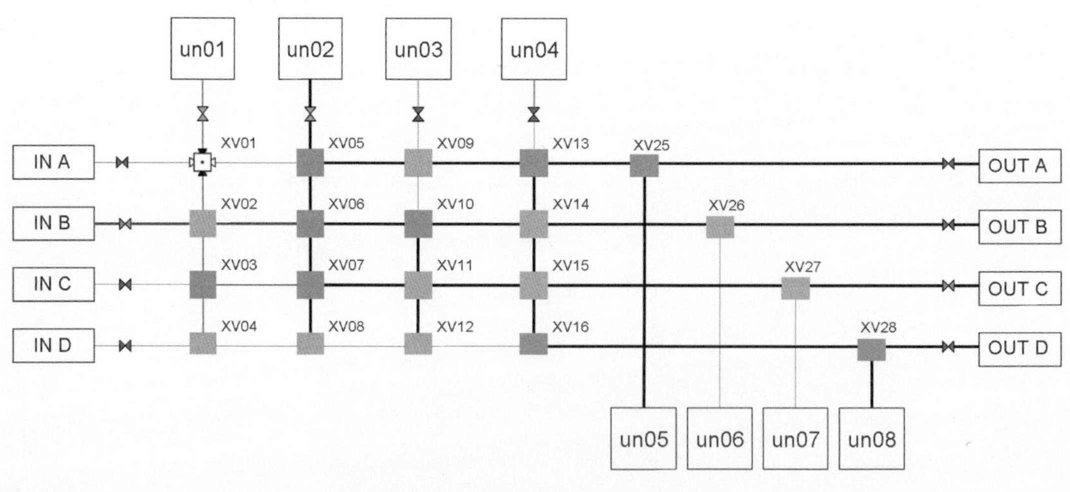

Figure 6.3.   Manifold with multi-port valves.

- "Unit-to-Unit coordination may be used to enable functions such as material transfers between units."

- "The control of the movement of batches may involve a number of choices between alternate paths. Although these choices may be made via links between Units, the Process Cell may also have to determine the routing."

So How Do Real Applications Handle This?

From looking at many projects, it is apparent that each project engineer has his or her own way of handling routing and transfers. Often, Phases deal with the equipment reservation and commands to set routing by treating routing as recipe parameter. (A route, by the way, is not a recipe parameter.) Often send and receive phases are also run simultaneously in the source and destination.

The Material Transfer Phases may be full of logic to handle the routing, or there is a phase for each route. That means that adding a new route requires adding a new phase or changing the phase logic. The result is lot of control software that is not flexible, especially when the plant changes.

Send and Receive Phase Coordination

Figure 6.4 is a good example of how send and receive phases are coordinated. It is important to note that the interface between source and destination is done by direct communications between the controllers—not via the batch manager. The sending and receiving phases use the values from inter-unit communication, and in addition, basic control such as exception handling use these values (e.g., if a valve fails the transfer stops immediately).

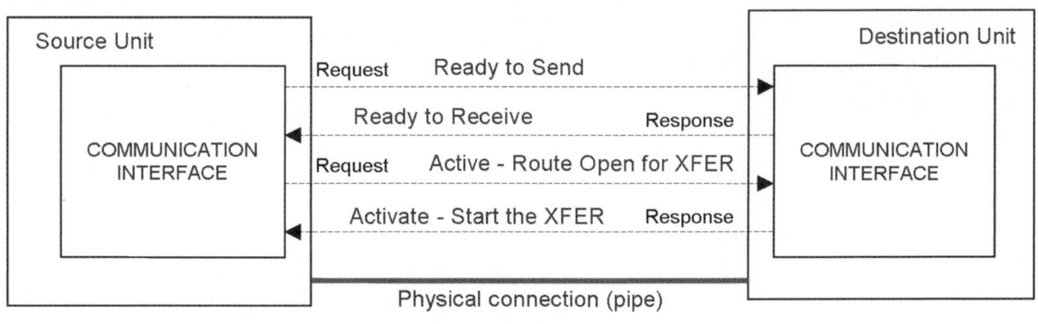

Figure 6.4.   Send and receive phase coordination.

How to Make the Transfer Routing Flexible

When a Web browser connects to a Web site, it does not care what happens in between. Maybe we can achieve something similar with transfers, so they are programmed using Generic Source and Destination objects rather than the specific equipment.

Recipes and Equipment Requirements

ISA-88.01 rightly makes the type of equipment to be used into a part of the Master Recipe. Then as the control recipe is executed, specific equipment that conforms with the required equipment is allocated (Fig. 6.5).

There may be piping or other paths between the source and destination. It helps to imagine a Transfer Object (Fig. 6.6).

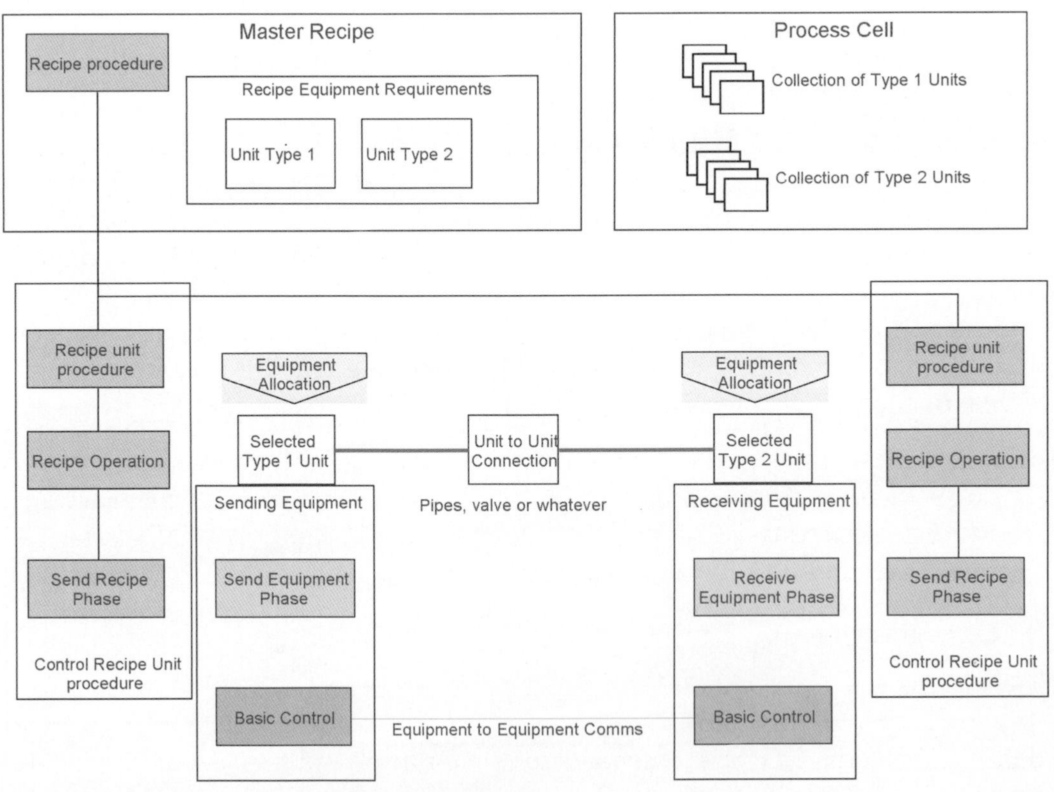

Figure 6.5.   Recipes and equipment requirements.

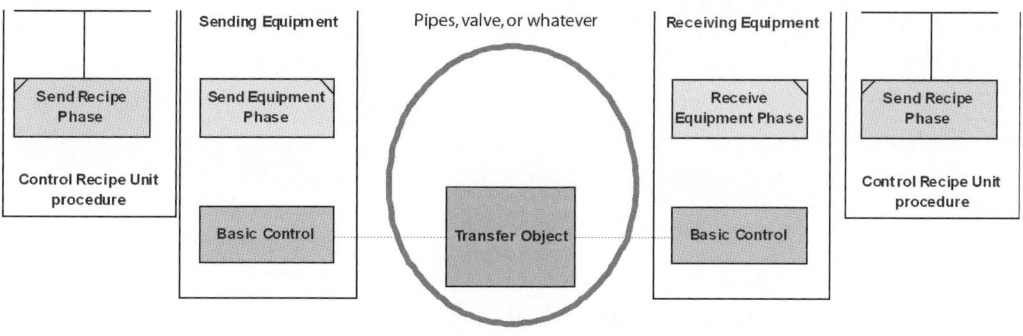

Figure 6.6.   The Transfer Object.

The Transfer Object

The Transfer Object is an entity that isolates upstream equipment from downstream equipment. The Transfer phases can be written as one sequence, referring to Source and Destination without caring which equipment is involved. Transfer Objects can be programmed to perform the following tasks:

1. Propagate Commands and States upstream and down or even across streams

2. Contain other transfer objects

3. Allow multiple transfers at the same time if the equipment supports it (shared use)

4. Contain entities that need status tracking (e.g., clean, dirty, sterile)

It is not difficult to program such things, and this can be done in Programmable Logic Controller (PLC) or Distributed Control System (DCS) code. This may well involve using a little bit of indexed addressing (e.g., where a destination number indexes a table so that a specific group of addresses for each destination are mapped to a single location). I think some try to do too much with their Recipe Managers because they underestimate what equipment control (i.e., PLC or DCS controller code) can do, and as a result they lose the power of modern control systems.

Conclusions

Controlling material transfers presents problems for flexible plants. ISA-88.01 provides little help. Phase logic is also not the best place to control transfer routes. Much better solutions are feasible, especially if you allow control code to do what it is good at.

Applying ISA-88: The Human Factor

Presented at the
WBF North American
Conference, Chicago, IL,
May 16–19, 2004, by

David A. Chappell
Batch Technology Manager
chappell.da@gmail.com
Procter & Gamble, 8256 Union Centre
Boulevard, West Chester, OH 45069, USA

Abstract

Automating manufacturing using ISA-88 concepts across all of the operational manufacturing boundaries has become the "standard" way of doing business. The current states of the technologies used in these automations require individuals with unique capabilities, or success is not always certain. Each operational area has its own unique needs and requires not only a high minimum capability in that area but also the ability to coordinate across areas.

This chapter explains what it takes to identify levels of capability and the benefits of taking advantage of this capability. What to look for while evaluating an organization's capability to deliver automation will also be discussed, as well as the results of not knowing your automation supplier's real capabilities.

People, Places, and Things

To fully understand the challenges faced in automating batch manufacturing and the relationships between ISA-88, the tools used, and the people who attempt this automation, a significant amount of background is required. My intent is for the material in this chapter to be useable by a wide audience, including people not familiar with automation or batch manufacturing. As such, I use many different analogies and go into detail that some may find "obvious" and others may consider "common sense." In relation to batch manufacturing automation, I often find it is only "obvious" when looking backward, and the good sense used is rare enough that calling it "common" is a misnomer. By the end of the chapter, you should be able to understand the magnitude of undertaking a batch automation project, why there are so many seemingly different automation choices, and what levels of capability to look for in the workforce required for a successful effort.

ISA-88.01: A Survival Guide through the Wilderness of Batch Manufacturing

Background

Until the internationally recognized ISA-88.01 standard on batch manufacturing was developed during the late eighties and early nineties, there was no clear common understanding of what batch manufacturing was really all about. Attempting to improve this form of manufacturing by automation was always a very chancy proposition. Even though corporations were internally successful in their custom automation efforts, they found them difficult to sustain. Reapplication was never a sure thing when their success depended on a handful of technology masters. These corporations operated in this new automation realm in ignorance of others and of the total environment in which they existed, much like the European explorers reaching the new continents during the fifteenth and sixteenth centuries. The breadth and depth of batch manufacturing automation can be likened to the breadth and depth of the North and South American continents that awaited those early explorers. Before the advent of ISA-88.01, batch manufacturing was an often confusing and difficult discipline to fully comprehend, and each attempted automation effort was like an expedition into the unknown from which many of those early automation explorers never returned.

What Is ISA-88.01 and What Is It Not?

ISA-88.01 is not a "How-To-Do-It" guide; it is also not a "What-To-Do" guide. It also does not attempt to address the needs of running a business. When, where, why, and how to manufacture a product is outside the scope of ISA-88.01 and addressed by other evolving standards. ISA-88 is not finished; parts .02 and .03 are available now, and others are in progress. ISA-88.01 is not easy or simple!

ISA-88.01 is a grouping of concepts and descriptions of all aspects of the batch manufacturing environment that can be used as a tool to help you find your way through the wilderness of batch manufacturing automation. Once you have established a path, it is up to you and whatever resources and tools you can muster to clear the path and make it suitable for travel. The tools are still somewhat crude and require a great deal of craftsmanship and mastery to use, much like the tools available to the early day explorers.

ISA-88.01 is also not a guide for vendor solutions; many vendors who traditionally service the manufacturing industry are claiming tools and products that address all of ISA-88. Be skeptical of such claims, and engage true masters of the technologies to evaluate them and advise you. If you do not have your own wilderness guides, find some with good credentials. The tools and approaches used to automate the business functions of a corporation are also not currently well suited for use in the automation of a manufacturing process. It is not reasonable today to expect to be able to use tools that manage payrolls, billing and ordering, and other business functions to effectively open and close real-world valves and run pumps to actually make a product. I do see the eventual integration of all the systems used in automation into a seamless, effective, and efficiently well-coordinated grouping of technologies that will allow a business to experience the level of total automation, similar to what is now being realized by the world's financial industry.

The ISA-88.01 activity model, as shown in Figure 7.1, is an excellent way to demonstrate how the activities of batch manufacturing fall into three distinct zones: Process, Operations, and Information. Each zone has its own unique characteristics, and each requires its own approach to automation. Each zone can be viewed as having over ten thousand different possible automation paths that one could use to traverse it. Just like in the real environment, you need tools to clear the paths and build the infrastructure to support transport and commerce. After you have explored and carved out your paths and built your infrastructures, you sometimes find that many of these paths are not something you would want to repeat and start to search for better ways to travel across the environment. The ISA-88.01 standard provides a way to reduce the number of possibilities that must be explored to discover your optimum output by eliminating at least the nine thousand or so "not-so-good-but-not-obviously-so" options. This still leaves you to sort through the thousand or so remaining options, of which nine hundred are "almost OK," ninety are "good," and only ten are truly "excellent." ISA-88.01 provides a ten-fold improvement over the past, and your probability of success is much greater! Even with this great reduction in possibilities, you will find that you

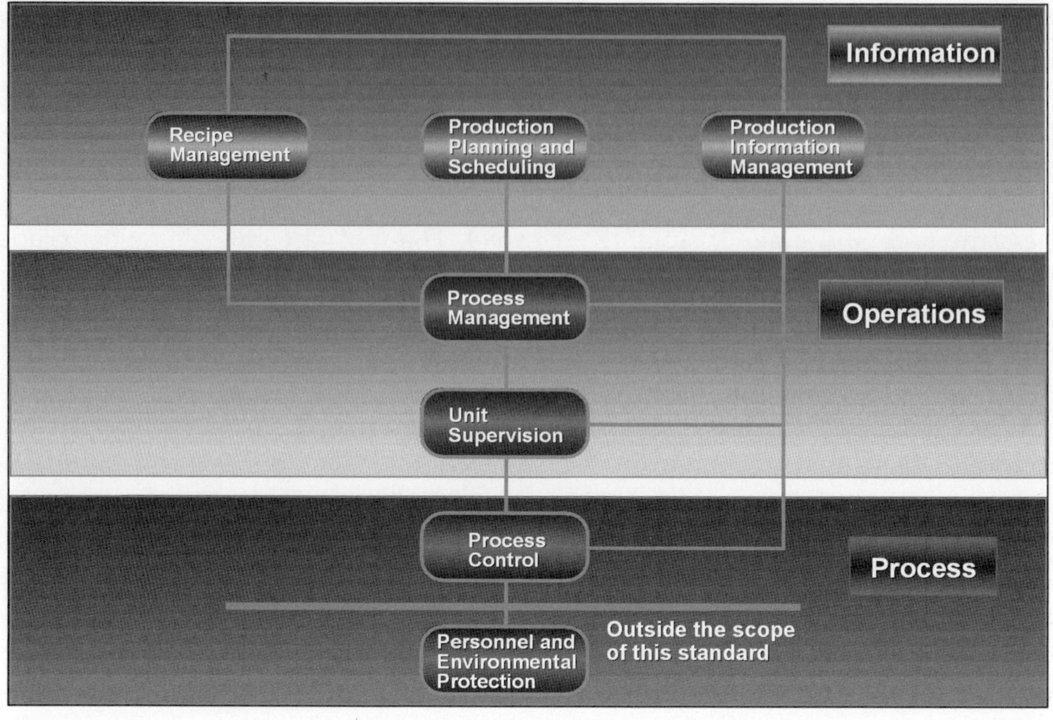

Figure 7.1. ISA-88 activity model and its zones.

can only sustain a few paths to explore, as each path requires significant effort and expense to develop and then support.

As the ISA-88 series of standards continue to evolve and the industry creates tools and technologies that are truly useful, those final thousand possible paths continue to be reduced to even more manageable numbers. But there is still much work to be done before those final ten "excellent" paths are identified, and vendors provide products that include the paths and required infrastructure as part of the product. When this finally happens, intrepid corporate explorers will no longer be required to make their own paths in search of the one that is best for them.

How Many Paths Does ISA-88 Offer and What Are the Vehicles Used to Travel Them?

Remember that each of the three zones (Information, Operations, and Process) has its own automation solutions, and what might seem to be an excellent solution for a given zone does not mean it will easily align with another seemingly excellent solution in a different zone. If your solutions do not support one another across the zone boundaries, then you have not truly found the optimum path for your corporate manufacturing automation needs. You will then have to continue your search, which can take decades because you do not really know if you have made a right choice until you have lived with that choice long enough to fully understand its strengths and weaknesses. Once you commit to a path, abandoning it and building another is an immense undertaking.

I'll share an analogy I like, which is represented in Figure 7.2. Each zone is immense: think of Colorado for Information, Arizona for Operations, and Louisiana for Process. Some people spend their entire career within a single zone covering immense distances and never crossing into another zone. The Process zone has chosen a surface road solution using gasoline-powered vehicles to move around the manufacturing business load. Operations has chosen a coal-powered rail solution, while information has chosen dirigibles using hydrogen gas (no smoking allowed). Each solution works very well within its own zone (except for the infrequent explosions of hydrogen-filled airships), and each domain has created an infrastructure to support its solution of choice. But when you cross a boundary into another zone, you must expend considerable effort to transfer your manufacturing business load from one transportation means to another, as the infrastructures do not allow free movement of your manufacturing business load between zones. You must also interact with the massive seafaring cargo

Figure 7.2.   Very large and very different areas.

vessel, which is the corporate business choice for moving the rest of the company's business load. These massive ships can only reach the edge of the manufacturing Information zone. They have no way to directly interact with anything else. Maybe we could learn something from the "container-shipping" industry! But we still have to address exploding airships right now.

In the Process zone, you generally find Programmable Controllers and intelligent instruments with Human Machine Interfaces (HMIs) that enable operators to carry out their mission in an efficient and cost-effective manner. Several engineering disciplines are involved in designing physical processes and their automation. Additionally, specialized technology-focused disciplines are required to implement the custom programming for the automation. The time horizon here is milliseconds to minutes.

In the Operations zone, you generally find process computers with dedicated applications and interfaces. These allow Manufacturing Operations (MO) to manage the process cells and units and apply procedures to make products and document production. Some tools and products are available for this zone, but there is still a significant amount of custom engineering and programming required that involve engineering and technology-focused disciplines. The time horizon here is seconds to hours.

The Information zone often uses Servers and Mainframe systems working with Corporate Manufacturing Execution Systems (MES) and Enterprise Resource Planning (ERP) applications that manage the flow of corporate data and information. There are many tools and products available for use in this zone, but custom engineering and programming are still present and involve engineering and technology-focused disciplines. The time horizon here is minutes to months.

Moving the business load across these zone boundaries requires significant custom engineering and programming that involves several different technologies and engineering disciplines. The practitioners of all these disciplines must work closely to enable success. Getting all the various suppliers of the technologies used in the different zones to work together to facilitate the movement of business data without all the custom effort is an area of great opportunity. These suppliers will eventually be forced to deliver more open products that enable this level of exchange. Until then, we can expect a hard sell for products that are best for the suppliers and not necessarily best for the end users. We can also expect a significant amount of "services" to make these products function optimally.

What to Do

Should you wait until the future catches up with your needs? Waiting until a technology is safe is a losing proposition. Your personal job may be safer here, but you are putting your company's future at risk. When a technology is safe to use in today's environment, you will find that others are already years ahead, leveraging it against you. If you have to play catch-up, it is much more costly to your business. Running a race with world-class track and field competitors and giving them a significant head start is not a wise choice. Placing as a top finisher is the only thing that is of benefit; starting the race after the winners have been decided is a questionable strategy.

Should you convince your business leaders to join you in working to identify and develop paths for your company that will most likely be where the future infrastructures will exist? What if you guess wrong? What if you choose poorly and find yourself stuck with approaches that are not compatible with the newly available commodity-level functions? This should not be a guess or a gamble but a known risk that is managed and shared. Identifying the boundary between the "bleeding edge" and the "leading edge" of any new concept or technology is difficult, and it should only be attempted by true masters. Trying to stay close to the leading edge will invariably result in a few minor cuts, but it is where the winners will be found. It is only by being aware of (and sometimes influencing) the leading edge of technology that a competitive advantage is realized—by being an early adopter of breakthrough technologies that will eventually become commodity products in the future. Those who prepare for and are part of creating the future will be taking manageable risks and will provide their companies a significant business advantage over those who choose to live in the past and react to the present. Those who attempt this without developing the proper level of mastery should really consider waiting for others to set the path and not partake in the race. If they attempt it anyway through either ignorance or arrogance, they are truly taking a "gamble" and are going to have to rely on "luck" in place of skill and capability. It is in this preparation for the future that the Human Factor comes into play: taking advantage of ISA-88.01.

Human Skills

All individuals develop "skills" throughout their lives. There are far too many "skill areas" for all people to be exposed to every one, so we must all pick and choose which ones are important to us, both in our personal and professional lives. Many skill areas depend on other skill areas, and when you have a large enough grouping of skill areas, it can be thought of as a "discipline." Mathematics provides a good example of a discipline. There are many basic skill areas, such as addition, subtraction, multiplication, and division. All higher-level mathematic disciplines, such as geometry and algebra, require these basic skills. As a person develops capability in a skill area, there are different levels of proficiency they can expect to achieve.

This is based on many factors. Some dominating factors are personal aptitude, personal attitude, training, technical mentoring, and experience. Continuing with the mathematics example: as students learn basic arithmetic, they receive training using textbooks and instruction from a teacher. Some students with the right "aptitude" will get the concepts very quickly and progress rapidly in their proficiency to apply that skill. Others will require more "mentoring" by the teacher to progress. Only by "experience" or "practice, practice, practice" will anyone truly learn a skill. Some skills require a lot more practice than others. Personal attitude (open and questioning versus arrogant and self-absorbed) plays a significant role in individual success or failure in this development process. It takes a great amount of mentoring to overcome the wrong attitude, and many mentors are not willing to invest the necessary energy and time. Proper mentoring can overcome aptitude and help individuals achieve their highest "level of capability." On the other hand, inadequate mentoring can provide people with the opportunity to learn many good lessons from their mistakes. This latter process is very ineffective as it only teaches "what not to do next time" without any advice on "what to do the next time." A proper balance between inadequate mentoring and overmentoring (dictatorship, actually) must be maintained, so as not to squash innovation.

There are many ways that these levels of capability can be measured. Grades in education are one: A for excellent, B for good, C for average, D for inadequate, and F for failing. What were your math grades? Another grading system that I find useful in industry is represented in the five categories or levels shown in Figure 7.3. These categories are beginner, novice, apprentice, expert, and master. As the pyramid suggests, there are always a lot more people at the lower levels than the upper. The common characteristics of each level are described in Table 7.1.

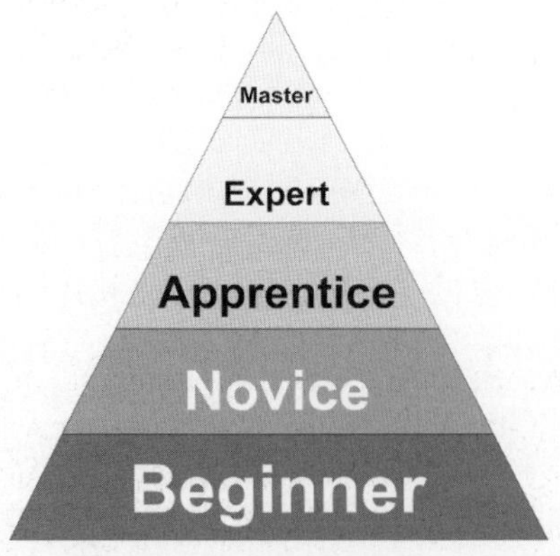

Figure 7.3.   Levels of capability and people concentration.

Table 7.1.	Common skills categories and characteristics
Skill level	*Characteristics*
Beginner	Beginners have the "book knowledge" of what the skill or discipline is about. They have little to no practical knowledge or experience, and they require constant attention. At this stage, they have an awareness of what they are about, with little to no real knowledge about what to do with their "book knowledge."
Novice	Novices have progressed to the point of general, real-world, practical knowledge about the skill or discipline. They usually feel they have mastered the technology, and they will proceed to apply it with marginal results. They generally do not know enough to understand what they do not know, and they require a great amount of direction from those who have a higher level of skill. If an individual never advances beyond this level, they often feel and act as if they have achieved a much higher level than they are entitled to, leading to confusion and failed endeavors. They are often energetic, enthusiastic, and fearless in the application of the skill or discipline.
Apprentice	Apprentices have progressed to the point where they have effective practical knowledge about the skill or discipline, and they can successfully apply that knowledge. Some people are quite happy to stay at this level, and they decide not to invest additional effort to advance. Others have no choice but to stay at this level, as they have reached the highest level that they are capable of achieving, and no amount of effort will enable them to advance. A few will progress on to the higher skill levels. They have respect for what they are doing, and they know what not to do and when to seek advice.
Expert	Experts have advanced beyond practical knowledge to deep understanding, and they can leverage their skills in ways that have never been thought of before. They are able to mentor the beginners and novices and provide resource help for the apprentices.
Master	Masters have acquired "wisdom" as well as understanding, and they are able to incorporate new learning back into the skill or discipline in such a way that novices and apprentices can make use of them. A master can advance a discipline and improve a technology. Masters must take care as they use their skill, so that those outside the discipline do not get a false impression as to the ease with which a discipline can be applied. Often those outside a discipline will expect master-level performances from apprentices and lower individuals.

If you have a sufficient capability level in a related skill or discipline, moving through the lower skill categories in a new skill area can require much less time. Take physics, for example. If you have achieved an expert level of capability in geometry and algebra, getting to an apprentice level in physics can be achieved much more quickly and successfully than if your starting point is a master level in government. While both individuals are excellent intellectually and are performing well overall, one will have an advantage over the other in developing capability in the new skill area of physics. This does not indicate the final level of capability each will achieve, just how quickly and with how much effort each will be able to move to the apprentice level. Then the practice, practice, practice starts.

As you can guess, it is very difficult and perhaps impossible to predict who will excel at a discipline, how long it will take them to advance, and who will and will not make it. When developing someone's skill, it is generally possible to "force" them to the apprentice level. Moving beyond that level requires desire and effort on the part of the individual and cannot be "forced" by any organization or person.

Technology Skills

As technology advances, new concepts and tools become available that can greatly increase the effectiveness of those who apply these technologies. These tools can also be used to artificially boost a person's perceived level of capability within a discipline. In this process, the technology acquires the ability to perform a skill that humans were required to do previously. This is a form of Product Encapsulation, in that a skill is embedded into a tool or technology and made available to all people, many of whom would otherwise not be able to successfully use it. This is not necessarily a bad thing, and it is perfectly OK to take advantage of it. Continuing with the mathematics example, we can take a look at the engineering disciplines, which rely heavily on high levels of mathematical capability. Not long ago, only a handful of people could successfully carry out the mathematics required to be an engineer. The opportunity to move up the capability levels in their engineering discipline was dependent on their proficiency at the required mathematics. This not only limited the number of engineers but also kept the number of experts and masters very low. The engineers of that day used the tools of their time: paper, pencil, slide rule (see Fig. 7.4 if you do not know what a slide rule is), as well as their acquired knowledge to meet the needs of the discipline.

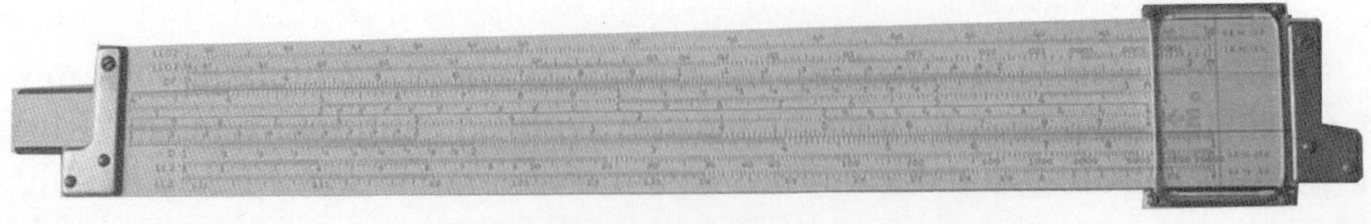

Figure 7.4.   Slide rule.

As technology advanced into the 1960s, computers became available to help with this, but the group still remained small. The computer technology required a whole different skill set (eventually leading to many entirely new disciplines), and many engineers found it more effective to stick with the tried and true slide rules. With the introduction of the pocket calculator in the 1970s (Fig. 7.5), the number of people capable of advancing through the engineering skill set increased significantly.

As pocket calculators increased in accessibility and performance in the 1990s (Fig. 7.6), the general public, novices, and beginners eventually became able to perform the functions once relegated only to a small group of experts.

Many of today's engineers are using calculators to perform complex functions, and some do not remember (if they ever truly knew) the detailed theories behind the functions the calculators perform. The end result is that they are successful in performing the function. Many will argue that this is not right, but it works. This evolution of technology-driven skill development took decades to unfold. Many different paths were traveled and abandoned before the industry identified the optimal paths (i.e., today's calculators) and created the products and tools that we now take for granted. A downside is that if the calculator batteries die, some will have to revert to their natural level or proficiency—at least until they can acquire new batteries. One must also be sure that the benefits of using a technology are not outweighed by the burden of having that technology. The early days of using computers in place of slide rules did not provide the right ratio of benefit versus burden. Once a function becomes part of a tool or product through Product Encapsulation, it becomes useable even at the beginner level of a given discipline. This will not make the beginners as capable as those who are more advanced, but it will narrow the gap between them.

The computer technology of the 1960s has evolved from a single discipline of programmers and database managers into many different products and tools with enough complexity and skill areas that many distinctly different disciplines have been created. These disciplines have become specialized, and they are focused on using their technologies to meet the needs defined by other engineering disciplines. This has created multiple engineering disciplines to satisfy a single need. Some examples of these technologies include Distributed Control Systems (DCS),

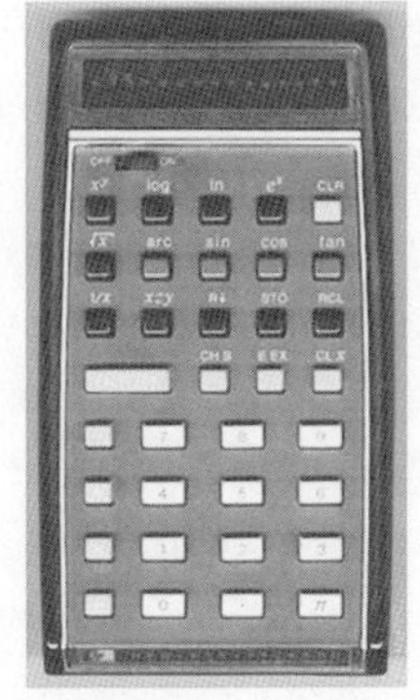

Figure 7.5. HP-35: The world's first pocket calculator.

Figure 7.6. The TI-83: A far more advanced calculator.

Programmable Logic Controllers (PLC), specialized Human Machine Interfaces (HMI), and MES to name just a few. As indicated in the Industrial Controller shown in Figure 7.7, these technologies have evolved to include many different functions.

Each function requires its own unique skill set. Just becoming expert in one skill does not qualify you in another skill. For instance, mastering the use of control loops in a controller does not mean you are competent in the application of advanced logic. Many people's entire career will focus on a single technology, and they will become highly competent in its application against requirements identified by others.

Who's Who, What Can They Do, and How Do You Know?

When the practitioners of a discipline have achieved the required level of capability, they understand and respect the complexities within that discipline or technology. If the proper level of capability is not present, then the potential for disaster is significantly increased (remember "learning what not to do next time"). Those who are dependant on that discipline for success are of little help because looking in on a discipline from the outside can often cause the true complexities to appear deceptively "simple and easy." What it takes to be successful is easy to misjudge and misunderstand without the ability to identify the proper skill set. This can and does lead to some very spectacular failures that often become excellent examples of what not to do next time (if you get a chance to have a next time).

Great care must be taken if you attempt to validate a discipline practitioner's true capability. There are many novices who successfully masquerade as masters.

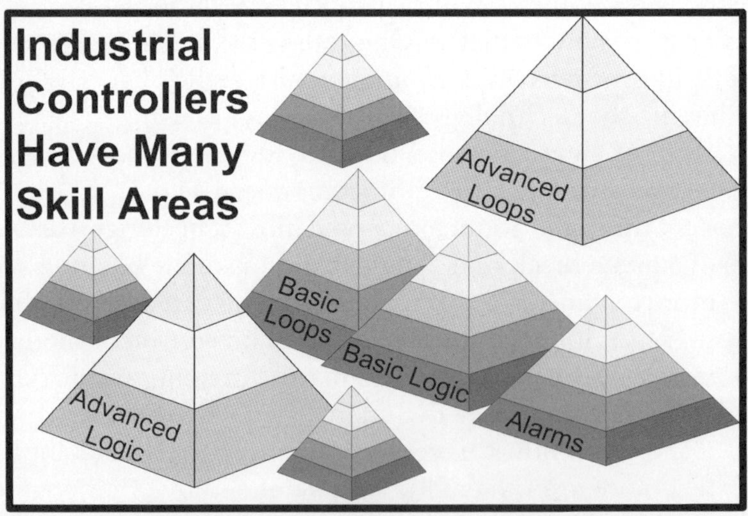

Figure 7.7.   Industrial controllers.

Remember what I said about looking in from the outside: you really cannot tell the difference between a novice and an expert unless you yourself have the required skill to make that determination. It is unfortunate, but often a person's advancement within a corporation is not necessarily due to capability. Advancement might occur because of time at grade or because the expert quit and the novice is the only one available. In these cases, the individuals do not have true knowledge or understanding of the technology and discipline, and without that there is neither respect nor fear. Thus these individuals often attempt things that are not practical. They do not have the scars to match the stripes awarded, and they become ineffective at meeting needs. But be careful—just because there are scars does not mean there is skill; a scar is not a sure sign of successful experience. Some scars may have been self-inflicted or received in unnecessary battles.

When looking at the general population of any group of practitioners, it is very difficult to identify what level a person is really at without firsthand knowledge. Individuals at different levels may look and sound similar to those who are not well versed in that technology or discipline. This is like the crowd in Figure 7.8: an individual's skill is not apparent.

Even when viewed by a peer, unless there is firsthand knowledge, it is difficult to be sure of one's skill level. For example, trying to identify an individual's driving ability is difficult without firsthand, in-depth knowledge. He might have made the trip from the West Coast to the East Coast in all types of weather, and he can provide you with documentation that he had no tickets or accidents. Does this mean you should entrust him with transporting your valuables or loved ones? You may find that during his trip, he only traveled in the day time, stayed off the expressways, stopped when the weather was bad, and caused others around him to have catastrophic crashes. Your requirement that he drives at night, in the rain and snow, and on the expressway may not be practical. You will have to ride

Figure 7.8.   An individual's skills can be lost in the crowd.

with him for a while to get the required sense of his skill level. It would be great if one could quickly identify a person's true capability level as indicated in the people pyramid of Figure 7.9. If we could do this, we could quickly know who to stay away from on the highway! Unfortunately, there is no magic method to see through the artificial measurement methods we have to rely on. Firsthand empirical data are required in order to actually know.

Manufacturing-Focused Disciplines (MFDs) and Technology-Focused Disciplines (TFDs): Apples and Oranges

When we look at the reasons ISA-88.01 came into existence, we must understand that it was created to help the engineering disciplines better satisfy a business need. This need was driven by the desire to reduce the cost of automation and improve the realized benefits to the end customers of manufacturing. Even though the creation of automated batch manufacturing systems was driven by engineering to improve the engineer's life, the ultimate beneficiaries of automated batch manufacturing systems are the business owners and operators. And as business owners and operators realized the benefits of these systems, it became a win-win situation for everyone. In order to meet the needs of the today's manufacturing

Figure 7.9.   A people pyramid showing skills.

business owner, one must first establish what their specific "detailed" needs are. This may require educating business owners about the direction the automation industry is taking. This is necessary because manufacturing's focus is not on applying technology for technology's sake (keeping engineers in business is not what manufacturers are there for). Instead, their focus is on competitively making products so that the company can make a profit. Keeping the manufacturing customer informed and making them part of the decision-making process is very important because they are the ones who really must request and accept something new. This is where the **"PLAN"** or "What-To-Do," is developed jointly between MO and the engineering disciplines that will be responsible for the actual **"AUTOMATION APPLICATION,"** which will reside within a given **"TECHNOLOGY."** With the current state of batch automation technology, there is still a great need for the corporate engineering disciplines to manage not only the specification and delivery of manufacturing automation, but also those who will actually do the delivery (programming and customization). The long-term ultimate goal is to have vendors create and supply products that can meet these business needs without requiring programming or customization, thus reducing the need for engineering disciplines almost to the point of nonexistence. Today's engineers should not worry about their jobs being in jeopardy because this process will take many years. After it is successful, there will always be other opportunities for engineers to apply their skills in finding solutions.

I have just identified two distinctly different disciplines that are involved in the automation of batch manufacturing. For the purpose of this chapter I will call them "Manufacturing-Focused Disciplines" (MFDs) and "Technology-Focused Disciplines" (TFDs). MFDs are disciplines like process engineering, controls engineering, and packaging engineering. TFDs are specialized disciplines that support the MFDs' efforts using tools (programmable controllers, computer applications, etc.) to create the automation required by manufacturing. Another way to look at the MFDs is to compare them to a symphonic orchestra conductor who will develop the score that the musicians (the TFDs) will use when playing their instruments. Then there is MO, which is the ultimate customer of all the automation effort. This is not really any different from the olden days I can still vaguely remember. The panel builders of yesteryear have been replaced by the TFDs of today.

Automation Development Process

MFDs must be able to understand the true needs of MO as well as the corporate political and physical environment. MFDs are where the Plan is developed that will be the basis for the automation created by the TFDs (Fig. 7.10).

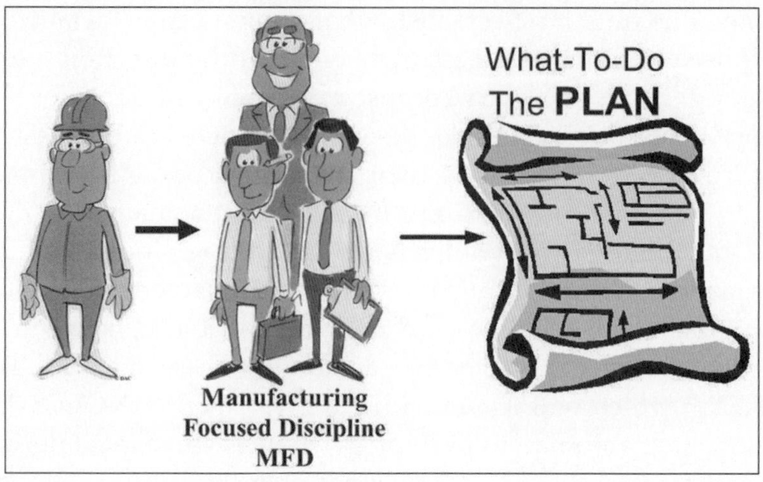

Figure 7.10.   MFD.

The Automation Application, the resultant work of the TFDs, is used by MO to improve manufacturing, as represented in Figure 7.11.

One very important step is to decide "How To Do It," often referred to in diagram form as a cloud with the phrase "Then A Miracle Occurs." This is the **"DESIGN"** that will be used by the TFDs in their work. The Design is a joint effort between MFDs and TFDs, with the right level of capability as represented in Figure 7.12. Figure 7.13 illustrates the "big picture" of these efforts.

Letting a TFD develop your Plan carries significant risk. If you must take this path, involve others who are knowledgeable in batch manufacturing automation

Figure 7.11.   TFD.

Figure 7.12. A miracle occurs.

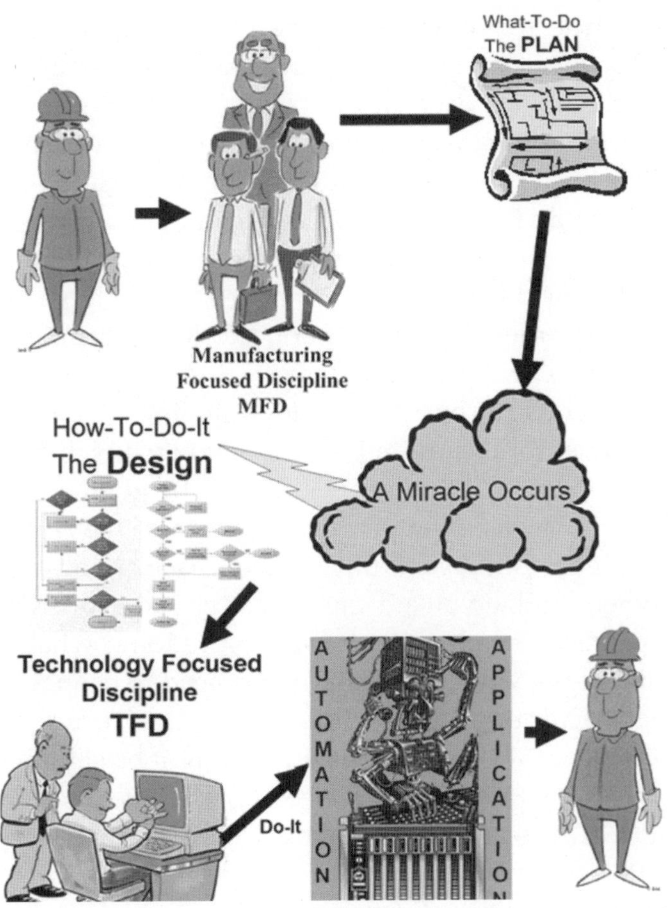

Figure 7.13. The big picture.

to help evaluate your Design. MFDs know your business environment, while TFDs know their technology. When TFDs are doing both the Plan and Design, it often leads to an Automation Application where your needs are modified to force fit them into the limitations of their technology.

The batch automation tools and technologies of today have not "**ENCAPSU-LATED**" enough of the functions necessary to meet all the automation needs of any one manufacturing zone, much less all three of them. What is available today is a box of tools that will be used to create the Automation Application described by the Plan and detailed in the Design. This resultant Automation Application will be used by MO in an attempt to run a more profitable business for a long time. Delivering the Automation Application is a costly effort, and replacement will not be done frequently, even if the Automation Application did not deliver what was really needed. Unless it is a total failure, MO will have to learn to live with it.

TFDs need to understand their technology and how to most efficiently lever-age it to meet the true requirements in the Plan, adequately staffing with the necessary capability levels to create the Automation Application. TFDs are often focused on the technologies used in the "Do-It" part of automation, with little regard to the underlying reasons defined in the Plan. TFDs are often contracted to do only the final part of a project—the Automation Application—using their skills with their products to do the delivery. MFD personnel will sometimes work with or for the TFDs to gain experience with the tools of that discipline. It is this group that will provide the effort to "Do-It," and hopefully they have the required capa-bility levels in the skills they are using. If not, excellent learning opportunities for what not to do next time will probably exist.

This process provides the opportunity for a significant amount of "creativity," as the industry searches for the optimum automation paths that will eventually be Encapsulated into the tools of tomorrow. This can be good, and it can be bad. If you are using your creativity to improve the Design of the Automation Applica-tion, this is good. Even if you do not fully succeed, each attempt will be better than the last. On the other hand, if you are using creativity to express yourself, meeting some deep-down artistic need, and you are just redoing what was done before differently, you are in the wrong profession and need to change to a more artistic discipline. Acting or sculpting might satisfy your desires better; although the com-pensation you receive until you become famous might be significantly less.

Some companies try to manage the need for creative work by capturing suc-cessful Plans and Designs and then reusing them in future works. This form of "**REPEATABLE PLAN**," sometimes referred to as a "CORPORATE PRODUCT," is not nearly as good as having it Encapsulated into a vendor's tool. However, it is much better than having to reinvent everything all over every time. A danger here is that the Repeatable Plan, as represented in Figure 7.14, may be very specific to a single solution and not work well in other situations.

This presents a real dilemma: How many Repeatable Plans can be supported? Should you force-fit many of your needs to a single approach? When is it right to unleash creativity and explore new solutions if the current ones are not an adequate fit? These balances are delicate, and decision making about these types of issues requires the highest levels of skill in the technologies.

This approach of using a Repeatable Plan becomes another tool built around specific needs and technologies. Both the MFDs and TFDs must develop the appropriate skills with this new tool to be successful in using it. If a Repeatable Plan does not become Encapsulated into a vendor's product, it will probably become one of those dead-end paths that work well for a while but could not be sustained over time.

MFD Practitioner Categories and Characteristics

Table 7.2 provides some insight to the different skill levels and capabilities of MFDs who work in the batch processing industries. As described earlier, an individual's journey through the different MFD capability levels is a personal one. You must accept that each person's path will be different, and there is a broad range of time it takes to move along the journey. Some will progress faster than others. You must also accept that not all who start this journey will be able to complete it.

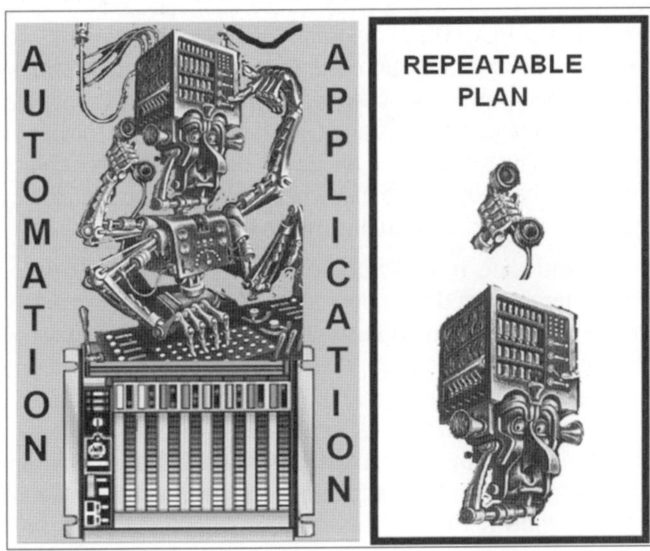

Figure 7.14.   A Repeatable Plan.

Table 7.2.	MFD characteristics
Skill level	*MFD practitioner categories and characteristics*
Beginner	Beginners can assist in delivery of a Plan based on guidance from someone who has achieved at least apprentice-level capabilities. They can help in the execution of a proven Plan by actual hands-on application of technology with guidance from others that have the appropriate levels of capability with that technology. They spend most of their time learning about their own discipline and the technologies used in its automation.
Novice	Novices can assist in managing the execution of a proven Plan. They are continuing to learn about the technologies used in the implementation of the Plan and they are continuing to develop their capability in those technologies as part of the Plan delivery. They have knowledge of the corporate landscape but do not yet know how to operate well within it. They are not capable of identifying if the automation required by the Plan is being properly created, and they are not ready to make these kinds of assessments.
Apprentice	Apprentices have progressed to the point where they can effectively develop all of the details required to carry out an already defined Plan. They have achieved "apprentice" level of at least one of the technologies used in delivering the Plan and have at least a "novice" level of capability with several others. They are capable of identifying if the automation required by the Plan is being properly created. They can manage those delivering the automation and also do a significant part of the delivery themselves. They can assist in developing a new Plan, but they are not yet ready to take the lead in the development of a new Plan. They can also assist in the development of a Design. They know the corporate landscape well and could operate within it if they were not spending all their time developing capability with technologies. People will often decide to leave a discipline during this stage, as they truly get an understanding of what a career in that discipline means. Most will use this as a logical point to move on in a career, with a few spending their entire career at this level and most never quite making it to the "expert" level.
Expert	Experts can create the Plan. They can guide the manufacturing business in identifying reasonable and beneficial automation using the available technologies defined by the corporate physical and political landscape. They are at least "apprentices" of several technologies used in their discipline, and they have at least a "novice" level knowledge of all others. In most disciplines, this often becomes a person's career choice if they make it to this level. These are the people who can work alongside the technology-focused experts to develop the Design required to actually implement the Plan.
Master	Masters are very old experts who have learned at least one discipline to the point that there are no unanswered questions. They have become an expert in at least one technology used in automation and usually several. They can make what is extremely difficult look easy. If they are not careful, they can give a false impression as to the true complexities of a discipline. They are capable of identifying a successful Plan that has reusable features. They can create a Corporate Product built around the technologies for repeatable application by those with lower capability in delivering against a similar Plan. A master can influence the suppliers of the technology to eventually Encapsulate features for the Corporate Product into vendor-supported tools. In some rare cases, the MFD master can create the Plan, develop the Design, and deliver the Automation Application without TFD assistance.

TFD Practitioner Categories and Characteristics

Table 7.3 provides some insight about the different skill levels and capabilities of TFDs who work in the automation industries. While there are some specialized tools specific to the batching industry, the products these specialized tools are part of are also used in other aspects of automation. Having advanced skills using the technology and tools in *other* automation areas does not indicate that a person has the required skills to be successful with applying the specialized batch tools. Every time a person uses a new tool for the first time, they start at the beginner level. They must earn the right to advance by demonstrating their skill and experience. As described earlier, a person's journey through the different capability levels is a personal one. You must accept that the range of time for this journey may vary and that not all who start this journey will be able to complete it.

One way to measure practitioners' capability levels is to determine how productive they are at using their product's Encapsulated features. The measure I like to use to help define the separation between categories is as follows: each level represents a 20% improvement in that person's ability to apply the product's Encapsulated features as the supplier of the technology intended. In this case, there is no creativity required—just use the technology as designed. Developing new functions using the tools available as part of a technology is a totally different effort, and those at the lower level are not prepared or qualified to undertake these types of applications. This does not mean they will not try—just that the results will most likely be another lesson about what not to do next time.

What Should Your Automation Team Look Like?

How many and what types of people are required to successfully automate your manufacturing? The answer is the always-dreaded "that depends." There are many factors that dictate your minimum requirements:

- How much automation?
- Which zones will be affected?
- What is the capability of your resources?
- Are you developing new functions just to identify a few questions that must be answered?

Earlier I identified three distinctly different zones: Process, Operations and Information. I also discussed the need to move between these zones. Each zone has its own Manufacturing Focus, and each zone requires many different TFDs to

Table 7.3.	TFD characteristics
Skill level	*TFD practitioner categories and characteristics*
Beginner	Beginners are people who have only been exposed to a technology and have little or no capability to use it. Some even think of the technology as magic, and at this level they will never be able to understand it but are content in leveraging the Encapsulated tool as is. They are not capable of using this technology to design a new application of any type. It takes a beginner 5 hours to do what a master can do in 1 hour using the Encapsulated tools of this technology.
Novice	Novices have progressed to the point of general knowledge, and with some guidance they can use the technology to implement a Design. Unless they are sufficiently experienced in another technology, novices will often mistakenly believe that they have mastered this technology at this point, and they will proceed to apply it as if they were true masters with marginal success, or as is often the case, spectacular failure. Given the proper mentoring, these opportunities will be a lesson in "what not to do next time," and with guidance from the higher level on "what to do next time," the individual's chance of successfully making it to the next level increases. Without proper guidance, the search for "what to do next time" will be costly and time consuming, as "what not to do next time" is often discovered the hard way. The inexperienced novices generally do not know enough to understand that there is a lot more to learn, and they require a great amount of direction from those who have a higher level of technical capability with this technology. People who have mastered other technologies have usually been humbled by these experiences, and they have proper respect for all technologies. They know there is always much more to learn than what appears to be obvious. This level can use the technology to create a simple custom Automation Application based upon an existing Design, but they are not capable of designing even a simple application. It takes a novice 4 hours to do what a master can do in 1 hour using the Encapsulated tools of this technology.
Apprentice	Apprentices can effectively apply a technology as it was intended using the Design. They can develop a Design for simple applications. They understand the technology to the point that they know when they need help and will seek assistance in a timely fashion for applications using unknown aspects of the technology. This group keeps the experts busy with questions and their need for help when using the technology to create functions that are not an inherent part of the tool itself. It takes an apprentice 3 hours to do what a master can do in 1 hour using the Encapsulated tools of this technology.
Expert	Experts are those who have developed enough expertise in a technology that they can create a complex Design based on the requirement of a Plan. They do not know all the answers, but they do know how to find them quickly or when to climb the mountain and meet with the master. They are capable of directing and training the novices, as well as providing resources for the apprentices. It takes an expert 2 hours to do what a master can do in 1 hour using the Encapsulated tools of this technology.
Master	Masters in the technology have learned it to the point that there are no unanswered questions. These people are few and generally difficult to locate. They often live lonely existences atop distant mountains as they work to design the next generation of technical products and tools and once in a while provide support to the experts. In some cases, a master technologist can create a Design and then do the Automation Application without assistance from the MFD. For this to succeed, the master technologist must have extensive experience in the manufacturing operations of the MFD that created the Plan.

address the automation needs within that zone, not to mention the two independent skill areas needed to move across the zones as represented in Figure 7.15. I do not mean to say that automation is impossible, just that it is not simple or easy, and when done well, it is extremely beneficial to manufacturing.

Totally automating a modern batch manufacturing facility is a task that can rarely be done by one team because of the deep level of capability across many different technologies that is required. You may have a single individual "responsible" for all aspects of the automation (usually an MFD practitioner), but the actual implementation will probably be done using many different teams. If you misjudge the work and do not staff properly, you may find yourself learning what not to do next time—if you are given the opportunity to try again. Often only one or at the most two zones are automated at any one time, greatly improving the possibility of a successful automation attempt. If you are developing a new Plan and using new technology to implement that Plan, the capability of both the MFD and TFD practitioners must be very high, or the risk of failure will also be very high. At the other extreme, if you are reapplying a Repeatable Plan with a known Design, reusing known technology and known practitioners, it is practical to use resources with much less capability.

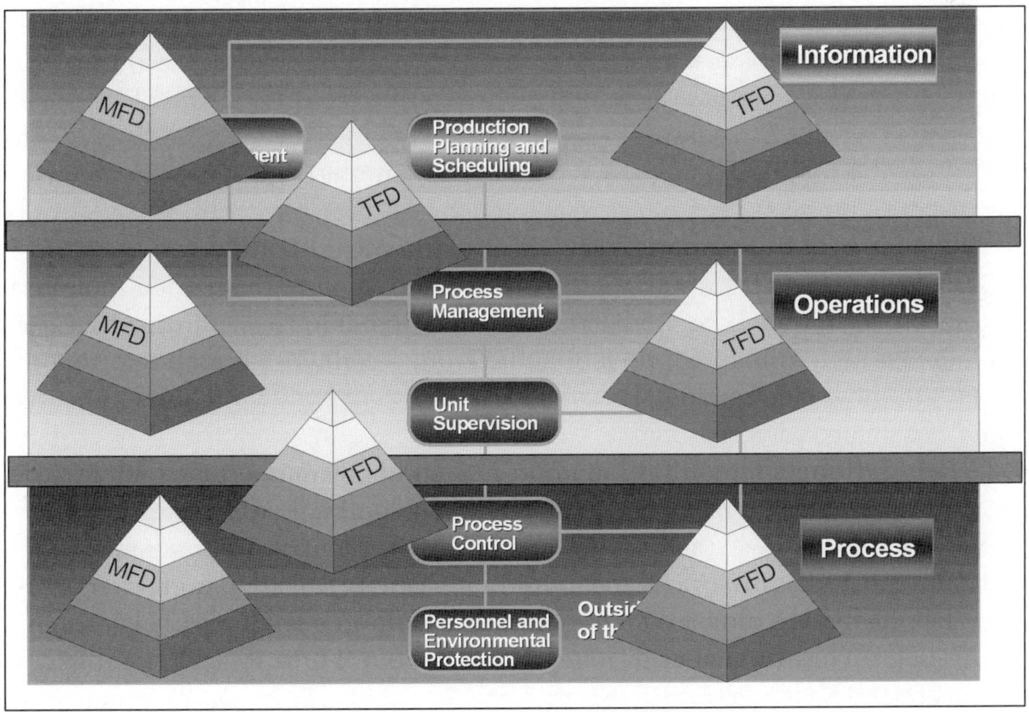

Figure 7.15.   Zones and skills.

While it is theoretically possible that one person or a very small team can do all this automation, it is very rare to find individuals with that level of talent. If you ever do, they will probably also have tight-fitting costumes and fight crime for fun in their spare time.

If you are fortunate and do have people this talented, make sure you have a succession plan in place and truly recognize your risk, should they become a causality in their crime-fighting hobby, or more likely, get bored doing automation work and move on to something else. In general, you must form a "team" of people with the proper skills to develop and deliver the required automation. When automating only one of the three zones, this "team" can be small. Automate more than one zone, and the team will grow significantly.

The MO customer really cannot distinguish the seemingly subtle differences between the people working in the different zones, much less the differences within a zone. In other words, the people and zones all look alike to the MO customer. This customer can become quite confused when dealing with the different groups, and care should be taken to make sure the MO customer is well educated about who is who and why it is the way it is.

Depending on the developmental needs of any automation effort, different pairings of groups as well as different skill levels will be required for a successful automation. It would be nice if every project could be staffed only by experts and masters, but this is not practical because of the scarcity of these resources. Cost would actually decrease using the higher-priced (but more capable) talent because of the increased effectiveness of these highly skilled individuals. This is nice when you can get it, but do not count on it. Depending on the requirements dictated by your project, you need to make sure at least the minimum capability levels of talent are working on that job.

If you are exploring a new path using new technology, your skill needs are going to be much higher than if you were reapplying a tried and true plan using known technology. The different mixes that I have seen work for the different type projects are shown in Figure 7.16. If you choose poorly, the results can be spectacular and not in a good way.

Figure 7.16 represents the four possible combinations of automation effort you can find in a single zone. As the developmental unknowns increase, the risks also will increase, as well as the requirement of staffing with adequate resources to address the developmental needs. When looking across all zones and the transitions between the zones, there are five different permutations of this table that must be taken into account. Care must be taken to ensure increased levels of capability in zones adjacent to one that is undertaking a significant development effort. Even though the zone under development has no direct impact on the adjacent zone and its internal use of its technology, the interaction between the zones will

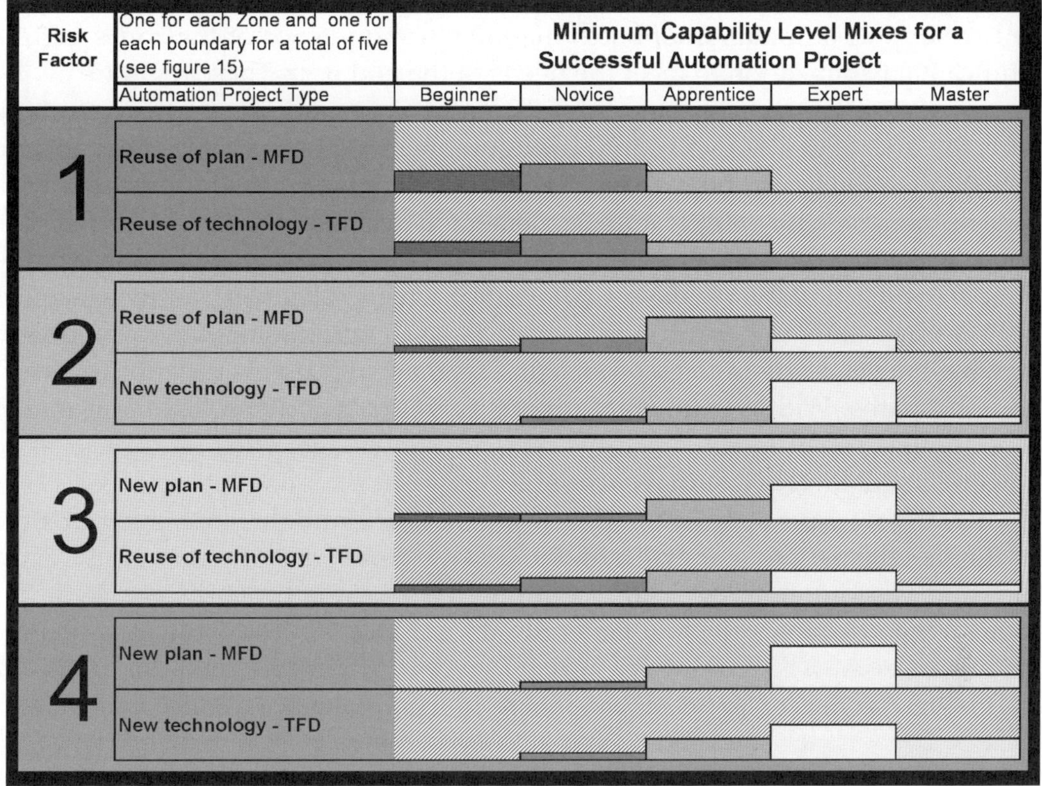

Figure 7.16.   Project types and capability requirements.

impact the adjacent zone and may require significant changes in the way the current technology is used to interact across the zone boundary. An apprentice most likely will not be able to handle necessary design changes within the adjacent zone to support the zone under development, so additional capability will be required.

Conclusions

The ISA-88 standard is not complete, and it will continue to evolve over many years. ISA-88 is not an answer, but it does eliminate a lot of the questions for batch manufacturing. The scope of ISA-88 is immense and really cannot be encompassed by a single set of automation professionals. Applying ISA-88 to automate a batch manufacturing operation is an adventure, with many paths to be charted and dead ends identified and marked so others do not make the same mistakes.

For the three zones of batch manufacturing there is no one solution. The tools used to support automation of batch manufacturing in each of the zones require significant customization to meet the needs of the end user. These tools are evolving, and just as the ISA-88 standard will continue to mature over the upcoming years, the tools will also evolve into configurable devices that will require little to no customization to meet the needs of the MO customer.

The people who design and implement batch manufacturing automation have many different specialties, and they must work as teams to define what to do, how to do it, and then make the automation a reality. Peoples' ability to acquire the necessary skills to perform the needed tasks in automation will vary greatly. Many will never really become proficient, and they will discover many "obvious" dead-end paths. Without sufficient capability yourself, it is impossible to determine another's skill level. Without firsthand knowledge, your best guess as to someone's skill level is just that—a guess.

Identifying your automation development needs and then matching them to the skills of the people who will provide that automation is of paramount importance and very much easier said than done. When done properly, you look like a hero. When done poorly, you generally do not look any way at all, as you are somewhere else.

ISA-88 has proven to be a great benefit to the automation industry.

Batch Data Analysis

Presented at the WBF
North American Conference,
Baltimore, MD,
April 30–May 3, 2007, by

Martin Zeller, Dipl-Ing
Bayer Technology Services
martin.zeller@bayertechnology.com
Bayer Technology Services, Building E 24,
D-41835 Dormagen, Germany

Abstract

Batch data analysis is comprised of three steps: data collection, data retrieval, and data analysis. In the first step, data collection, all potentially relevant process data have to be collected. To start analyzing a specific situation, the data relevant for this situation need to be selected in the second step, data retrieval. To yield information, these data have to be analyzed with process knowledge in the third step, data analysis. The concluded information is then used to develop process improvements.

Data collection is usually the easiest step, especially since the introduction of standards like Object Linking and Embedding for Process Control (OPC). Nevertheless, pitfalls are possible. To analyze batch processes, all batch events as defined by ISA-88.01 and their time stamps should be collected.

Data retrieval (e.g., all measured values of a dosing step for several batches or trends for selected tags during a recipe operation of interest) is only possible if the process data and the accompanying batch data are recorded and easily accessible.

Data analysis needs "know-how" in process technology. Root cause analysis can be supported with statistical methods. The selection of methods and the interpretation of the results are carried out by process engineers and supported with appropriate tools.

These three aspects of batch data analysis are carried out in several processes at Bayer AG. This chapter will show Bayer AG's general approach and results from individual projects. In all projects, the key success factor was the combination of IT know-how for the appropriate setup of systems and the process know-how for the beneficial application of these systems.

Introduction

The design and continuous improvement of efficient technical processes are major drivers of business success. Success factors include reduced waste, decreased cycle times, decreased consumption of resources (i.e., energy and raw materials), and increased throughput and manufacturing of products.

To get the best results, it is not sufficient to do Key Performance Indicator (KPI) optimization with standardized KPIs while individualized aspects of the process are neglected. It is necessary to have the relevant engineering know-how to gain insight from valid, up-to-date data and perform improvements thusly. This chapter will consider both the accessibility and presentation of process data and the way these data can be interpreted by engineers. I will argue that the presentation of data should be divided into standard reports that allow for individual, flexible online data analysis.

Data Collection

Data need to be gathered at their source, which is the technical process controlled by a Programmable Logic Controller (PLC) or Distributed Control System (DCS). These systems bring together all measured values from and all outputs to field devices (i.e., sensors and actuators). These systems do not archive data over long periods of time. Historian data systems or Process Information Management System (PIMS) are used to archive these continuously flowing data. Other types of data contain order- and batch-related information. These are derived from batch control systems, from manual input, or from Enterprise Resource Planning (ERP) systems. Batch-related data may also stem from a laboratory with or without using Laboratory Information Management Systems (LIMS).

All these systems are accessible through either OPC or SQL. Proprietary interfaces are not used now but may be necessary to access data from "old" systems. Nevertheless data access using industry-standard or proprietary interfaces is straightforward. Storage of measured values (i.e., trend data) is usually performed using dedicated archives and compression algorithms like the well-known "swinging door" algorithm. Batch-related data are usually stored using relational databases. Their proper design is critical for the retrieval of data. Only a well-designed data model allows for fast data storage and data access.

The most important aspect of data storage is data security. First of all, the transfer of data from the PLC or DCS to the archive must encompass all data; data loss is not permissible. Most DCSs offer a short-term data archive. Using this archive and historical data access (e.g., OPC Historical Data Access [HDA]), the long-term archive of the historian can be sourced safely using a buffered interface. Data from a PLC without short-term storage need to be transferred with a highly reliable technology (e.g., using redundancy).

Data Retrieval

The data model for the storage of batch-related data has to comply with standards, especially ISA-88.01. There are two major components: the equipment model and the process model. Most data are derived from the phase level of the process model. These data include run times, quality measures, operator comments, dosing values, and produced quantities, together with material IDs and lot IDs.

At Bayer AG, we developed an appropriate data model and have used this in combination with an SQL-RDBMS for more than ten years now. Of course the model was improved over time, but the basic concepts from ISA-88.01 (and formerly NAMUR NE33) remain unchanged. Not all data sources comply with the data model. In these instances, the data are rearranged for storage. Not all plants or projects encompass all the levels of detail offered by ISA-88.01. Appropriate adaptation of structures and terminology is in all cases necessary (since most plants use a historically developed nomenclature). We never experienced severe deviations from the general data model and were always successful in mapping the data to our data model. This is most likely due to the "invariant" nature of batch processes and the widespread usage of ISA-88.01.

Data Presentation

There are two major data user groups: personnel using preconfigured standard reports to run their daily business and expert users doing detailed data analysis to solve individual requirements.

Preconfigured Standard Reports

Reports are usually standardized within a plant to focus on specific periods (e.g., shift, day, week) and on batches and campaigns. Users of these reports include operators, shift foremen, plant managers, and the like. These reports typically reflect the structure of the control recipe producing the data shown, and they are of course aligned with ISA-88.01. Preconfigured reports can be set up according to ISA-88.01 independent of the individual recipe to fit any ISA-88.01 compliant recipe data.

Detailed Data Analysis

Detailed data analysis is usually done interactively. To help non-computer specialists perform data analysis, the information must be accessible by a few mouse clicks. This can be achieved by creating a comfortable user interface offering tree views for easy navigation and context-sensitive menus.

Single and Multiple Batches and Statistics

Data analysis for a single batch is similar to data analysis for batch reports and carried out according to ISA-88.01 structures. Multiple batches can be compared using run times, material consumed and produced, or other technical KPIs. These figures indicate deviations and can be used for Statistical Process Control (SPC). More detailed analyses of deviations can be performed by analyzing the trends of a process value over time on a relative time axis that begins at the start of the considered object or operation. Process values and combinations thereof that reflect the behavior of the technical process can be used as KPIs to give fast insight. Fingerprints, a combination of several KPIs, help us examine multiple batches quickly.

"Engineering" Driven View

Experts can exert multivariate statistics (e.g., Principle Component Analysis [PCA]) to gain insight and perform root cause analysis. Experts with process know-how will use the data to get information about the technical process. They can carry out case studies and perform root cause analysis for troubleshooting and process improvement. As long as the data are easily accessible, experts do not need to be computer specialists to access the data.

Summary

Data are a valuable asset. To unleash their potential, data are to be stored, retrieved, and interpreted appropriately for systematic process analysis and sustainable performance improvement.

Control Room Information for Batch Processes

Presented at the WBF
European Conference,
Mechelin, Belgium,
October 11–13, 2004, by

Milton Crofts
Milton_Crofts@afchemicals.com
Aroma and Fragrance Chemicals
(AFC), Widnes, UK

Abstract

The way information is displayed in control rooms in the process industry has developed a long way from the original gauges, chart recorders, and lamps to the sophisticated Graphical User Interfaces (GUI) of today. But does the increased ability to acquire and display more data mean that better information is being conveyed? This chapter describes how modern technology has not been used to its full potential in control rooms. A methodology is then developed to make use of the available technology in order to assist in the delivery of information to control room operators. In particular, this chapter addresses the problems in the presentation of plant data in the context of batch processes.

A Brief History of Data Acquisition and Display

From the early 1980s, developments in computer technology brought down the costs of computer displays and made them widely available. Primitive graphical representations of processes started to become available around this time, particularly with the introduction of DOS. In the mid-eighties, a combination of PC-based computing, color graphics, and the introduction of Programmable Logic Controllers (PLCs) gave rise to cost-effective data acquisition and display capability, which suddenly became available across the process industry.

The DOS-based displays were limited in the amount of data that could be displayed. This limitation forced designers to think carefully about how displays were configured and arranged. Usually, there would be a series of overview displays from which lower-level plant and instrumentation type displays could be accessed. The manipulation of data into information at the display level was difficult, however, and the display of raw data predominated. There was no real consensus about the best practice regarding the layout of displays, nor was there any standardization of color schemes. Only a handful of colors were available, but even so color schemes tended to vary from company to company.

Apart from the ambiguity of information conveyed through color, there was also an overuse of alarm data. The developments in Distributed Control System (DCS) and PLC technology meant not only that more data could be acquired but also that a series of attributes, including alarms, could now be associated with each signal. Those involved in the configuration of these plant control systems tended to set alarm limits on all signals, often at quite arbitrary levels. Rarely was thought given to the kind of action the operator may need to take on receiving such alarms. The assumption, if only a subconscious one, was that more equals better. The result, however, was that a large number of alarms were raised on some kind of alarm banner that held no meaning whatsoever to the operator. Worse still, where useful alarms did exist they tended to be buried within a stream of irrelevant data.

The absence of standardization or even of a clear idea of best practice was further complicated by the introduction of Windows-based displays at the end of the eighties. Of course, Windows and Microsoft in particular brought much-needed standardization to control systems hardware, software, and networking. But the increased flexibility in the presentation of data caused other problems.

Now there is, in practice, an infinite number of colors available, multiple windows, three-dimensional and animation effects, video imaging, and much more functionality. When asking a designer why he employs a particular technique in the configuration of a graphic, he is quite likely to reply "because I can." Glancing at the marketing documentation of any control systems vendor, one can find an abundance of descriptions of all of the design functionality described previously and more brought together in multicolored and multifaceted displays. But have all these developments in technology necessarily improved the efficiency of conveying information to the process operator? For example, does a sophisticated three-dimensional representation of a valve convey any more information than a simple ISA symbol? These questions will be addressed in the following sections.

The Process Operator

The role of the operator has changed in many instances over the years, from one of active engagement with the process on a manual plant to one of relatively passive monitoring of an automated plant. This is not to say that the role of the process operator has been diminished. Indeed, with the enhancements in the ability to acquire and to display data, the operator has more potential than ever to keep processes in control. Operator response to an abnormal situation is a key factor in loss prevention and therefore in the profitability of the company.

However, this abundance of data and the lack of useful alarm information have meant that, in practice, the operator can still only operate a small number of processes. This is because the operator must monitor large amounts of data in order to determine the current status of the processes for which he is responsible. Clearly the operator can only assimilate so much data, and it is this that limits his scope of responsibility. In a situation where operators are deployed to be dedicated to a particular process, it is likely that they will have to perform both control room and field operations, and this can further undermine the ability to use limited resources efficiently. In Figure 9.1, the changes in the level of

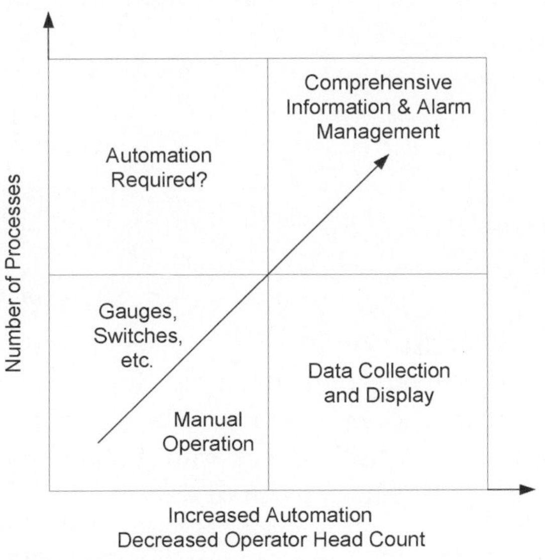

Figure 9.1.   Process operator versus automation.

information required is shown with respect to the number of process operators per process and the level of automation.

The technology now available has the potential to deliver information to the operator in a far more efficient manner than in the past, but this implies a significant change in the mode of operation. In other words, pressure to reduce operator headcount has meant that new modes of operation need to be found, and this is forcing changes to the way that plant data are displayed. The best practice in the industry points to using dedicated control room and field operators but also recommends rotating operators between processes and between control room and field operations.[1] In this mode of operation, the control room operators would need to monitor many more processes than on the traditional site. To achieve this, mechanisms need to be found to deliver condensed information to the operator rather than having the operator scan data.

A Methodology for Improved Delivery of Information

With modern plant control systems there is enormous flexibility in how data can be presented. In practice, the tools are now available for the designer to achieve almost any display format he can imagine. Add to this the sophisticated alarm-handling functionality that is generally available and it should be possible to deliver information reliably and in a timely fashion to the control room operator.

The methodology proposed here seeks to apply the technology available and the current thinking in best practice in order to address the following issues:

- Display color
- Display hierarchy
- Overview displays
- Operation displays
- Alarm handling

The interest here is focused on batch processes. The proposed methodology assumes that the processes in question are highly automated and are controlled by use of a modern batch management system. Normal operation should entail little operator intervention in the control room apart from having to respond to various prompts to select plant equipment or to input data.

Display Color

A consensus is emerging in the industry about best practice in the use of color for operator displays. In general, a cool screen approach should be adopted. The cool screen avoids the use of bright colors but utilizes grays, blues, and greens for normal operating conditions.

Abnormal situations should be brought to the attention of the operator by the use of small areas of bright color, typically yellow or red. It is possible to provide additional focus by flashing an unacknowledged alert.[2]

With batch processes, pattern recognition seems to play an important role, not just to identify plant status, but also to assist in the awareness of sequencing. For this, changing the distribution of color is a very effective method to inform the operator precisely which stage of a process is active.

Contemporary discussion often focuses on the ergonomics of control rooms, and some of the issues surrounding this are brought into current thinking on use of display colors. For example, it is recommended that design displays to be of low contrast in order to avoid eye strain. In addition, a light background is recommended to help optimize the overall lighting effects of the control room. Although it is the case that displays designed along these lines can be pleasing to the eye, care should be taken not to lose important information. For example, there may be several identical processing units, in which case prominent labeling would be required in order to avoid confusion and mistakes. Also, it has been found desirable to be able to identify process variables quickly on a display. Both of these examples require high-contrast areas within displays. Other objects may be of low contrast. For example, level or temperature switches can blend into the background unless in the alarm state.

It is clear, then, that it is necessary to have a balanced outlook with respect to the use of color. On the one hand, take into account current guidelines. On the other hand, never lose sight that the overall objective is to convey information.

Display Hierarchy

The objective is to develop a display structure that facilitates an expansion of the scope of operator responsibility. To achieve this, the number of displays per process must be reduced as the operator can only assimilate information from a limited number of screens. This suggests a layered, hierarchical approach whereby detail is increased toward the lower levels of the structure.

In principle, the hierarchy can have many layers, from site overview down to individual device. The number of layers, however, is dependent on the individual application and on the type of process. Reporting by exception should be employed as much as possible, whereby the operator should only be alerted to a particular process when a problem is occurring. Batch processes, however, have the added complexity of sequencing and the need for the operator to know which stage of the process is currently active. A control room operator may very well be coordinating a team of field operators, so field operations such as sampling need to be scheduled. To achieve this, the control room operator needs to be aware of the sequence at the unit level, and this effectively sets the top layer of the display hierarchy for the operator workstations in batch processes.

For batch processes, there may very well be higher-level displays that are appropriate for supervisors and production management. In addition, a site overview may be useful to show which plants are active and to indicate whether or not they are healthy. A technique that is sometimes adopted is to display a site overview on a large wall-mounted screen. In the following section, we will develop a methodology with respect to the operator workstation and as such use the unit process as the highest-display level. The methodology can be easily expanded to a multilayered structure using the same principles.

Overview Displays

An overview display for a reactor unit process is shown in Figure 9.2. An overview should be abstract in nature and assist in providing a conceptual view of the process, similar to that of a process flow diagram. It is important to note that the overviews generally need to be process specific. In the author's experience, an attempt to build a general reactor overview tends to clutter the available space with information that is irrelevant to the process in question. A reactor such as the one shown may very well be a multipurpose unit used for a number of processes, utilizing a range of associated vessels and services. However, a separate overview should be configured for each process rather than attempting to display everything in a single graphic. This approach undoubtedly adds to the engineering effort required for the configuration of the displays, but it helps clarify the information.

The various objects on the display are ordered to reflect the process sequence and to indicate the various services available to the reactor, such as temperature control and vacuum system. A banner is provided at the top of the display to give process description, summary batch information, and some key process variables. In a typical overview display, the color scheme is simple: gray is inactive or normal, green is active, yellow (or yellow flashing) represents failure or alarm, and different shades of gray are used for different objects containing summary information. The colors of lines simply indicate whether or not the line is completely

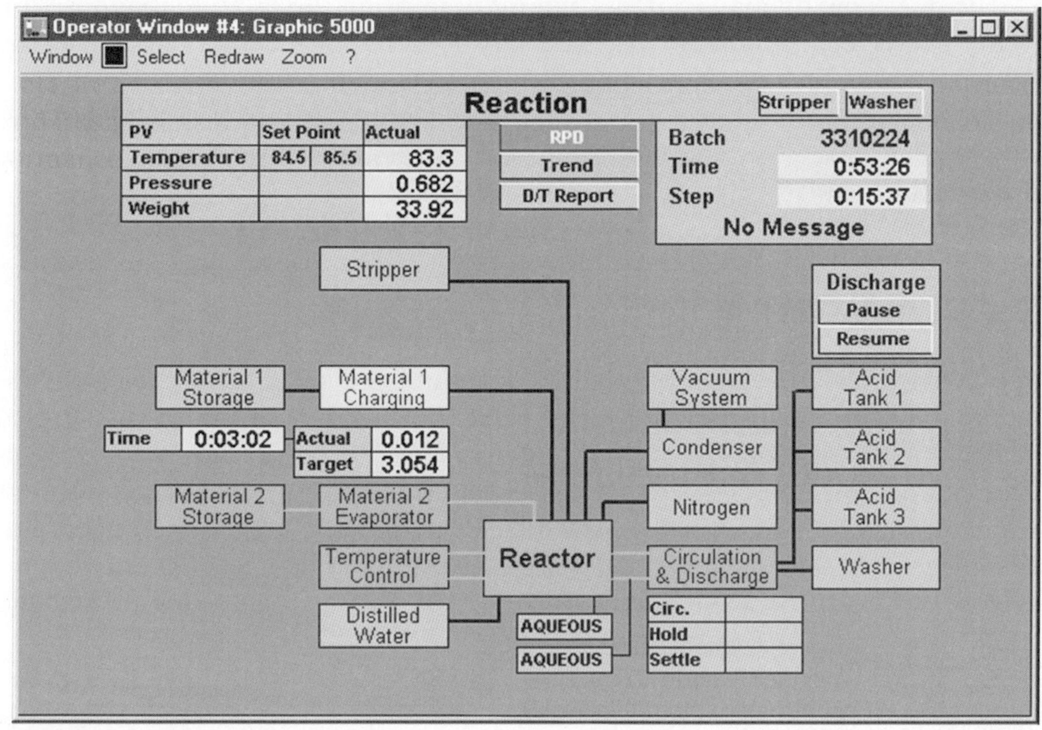

Figure 9.2.   Overview display.

open, using green for open and black for not open. For additional clarity, it is useful for operation-specific information to be visible only when the particular operation is active. In Figure 9.2, the target and actual charge weights for Material 1 are shown along with the time taken. This information disappears when the charge is complete.

Each block on the display contains summary information pertaining to the underlying operation. For example, the block labeled "Material 1" contains summary information regarding the charging of Material 1 to the reactor. At this overview level, the information should be the result of a comparison between the intention and the actual outcome. When there is no request to charge, then the block is deemed to be inactive and is color-coded gray. When a request for the charge is received, the block flashes yellow until the line is fully open, at which point the block turns green to indicate it is active. If at any time during the charge the outcome no longer matches the intention, the block is deemed failed or in alarm and flashes yellow. If the problem that has caused the charge to halt is due to one of the services, then that block will also flash yellow.

The overview display contains summary information only. To diagnose the specific nature of a problem it is necessary to access a lower-level, operation-oriented display. This is achieved by clicking on the block of interest. Operation-oriented displays are described in the next section. It should be noted that the delivery of abnormal situation information is assisted by use of the system alarm handler. This facility is also described in a later section.

Operation Displays

Operation-oriented displays represent the layer in the display hierarchy below the process unit overview. These displays provide a detailed view of a particular process operation or service. The display shown in Figure 9.3 is used for the operation to transfer the contents of the stripper vessel to the reactor and may be accessed from the stripper block on the overview. The display is similar to a plant and instrumentation diagram in that it shows objects down to the equipment level. It differs from a Piping and Instrument Diagram (P&ID) because instead of focusing on a single

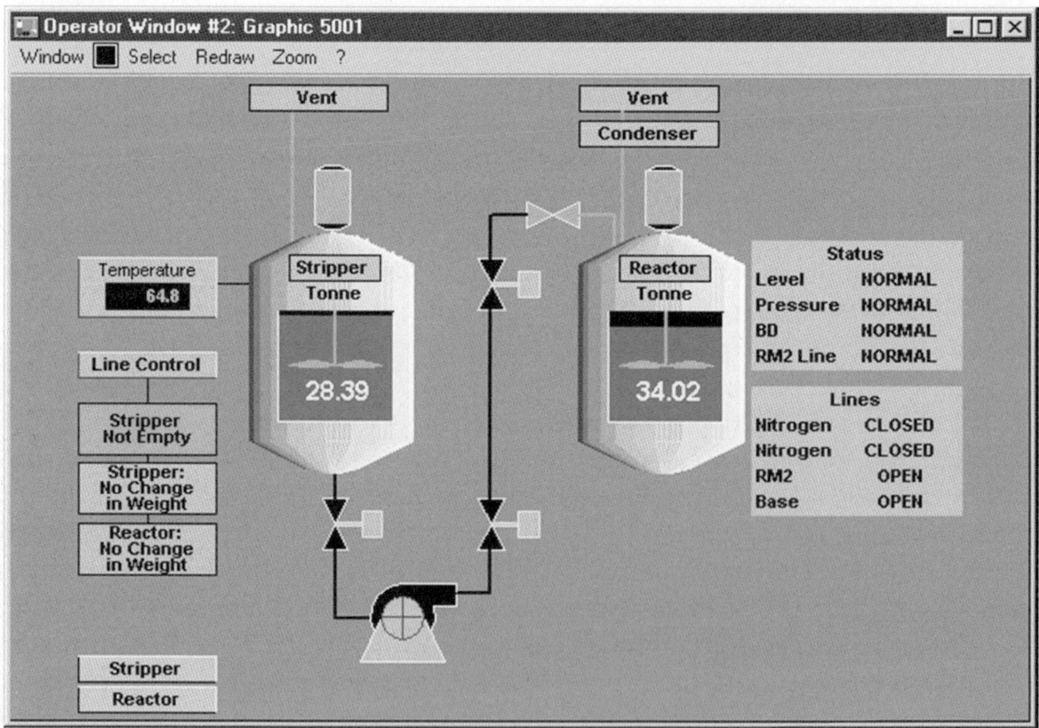

Figure 9.3.   Operation display.

processing unit, it displays the relationship between two processing units. It shows all the plant and instrumentation that is required to transfer the contents of the stripper to the reactor. The color scheme for the typical operation display is similar to that of the overview display: green for valve open, pump running, and line open; black for valve closed, pump stopped, and line closed; and yellow for failed. For status switches, gray represents normal and flashing yellow indicates an alarm. Process variables are displayed as green values on black backgrounds. These may also flash in alarm if configured to do so.

Often there are other device attributes apart from open or closed that should be made visible at a glance, such as the following:

- Device interlocked

- Field or DCS mode selection

- Multistate device

In addition, there are various other attributes that one might desire to be readily visible, but in general it is a good idea to present these attributes as shaded areas of the main device symbol. In the example given, an interlock is shown as a light gray shaded region of the pump or valve.

The level of detail on the operation display should be such that a quick diagnosis of any problem is possible.[3] Minimally, any interlock that could prevent the operation from taking place should be shown. Where it would be too cumbersome to show a particular interlock (e.g., because it was part of another subsystem), a reference may be made to it by way of a block similar to those used on the overview. In Figure 9.3, references are made to the vents of both vessels and to the condenser of the reactor. These are shown as blocks at the top of the screen. These blocks imply that there is one or more aspect of these subsystems that may cause the transfer operation to fail. A high vent line temperature would be a typical example of this and would be indicated by the block flashing yellow on the overview.

Other supporting information may be incorporated. In the example illustrated in Figure 9.3, the transfer is complete when the stripper vessel is empty. The test to determine that the stripper is empty is indicated by the three blocks on the left-hand side of the display. During a transfer, these three blocks would be green and would change to gray when the transfer is complete.

Under normal operating conditions the display should exhibit a cool effect, as shown in Figure 9.3. Flashing objects indicative of problems would then quickly attract the attention of the operator. Ideally, when the operation is inactive, the display should remain cool. However, ensuring that nothing flashes when the operation is inactive can take a great deal of effort. Careful consideration should also be given to startup or maintenance conditions in order to avoid spurious alarm information.

It is important to avoid making the displays too complicated. If detailed information about a particular object is required, then this information should be made accessible by clicking on the object in question. With Visual Basic for Applications now embedded in the display software of many control systems, it is possible to exercise a great deal of control over the kind of information that is accessible. For example, it is possible to click on an object to pull down a menu that provides access to drawings, specifications, Webcam, and so on, as well as a faceplate for operational purposes. Links to other displays or other data sources are useful, but these should not clutter the display. If a large number of links are required, then they could also be accessed from a drop-down menu. Three-dimensional objects should be used with care. If a three-dimensional view of a valve is required, for example, then provide a link to a drawing in a drop-down menu and keep the valve symbol simple on the display.

Alarm Handling

In order to begin this section, it is first necessary to define what is meant by an alarm: "An alarm is advice to the operator that immediate action is required."[4]

There has been great interest in the management of abnormal situations in recent years, particularly since the Milford Haven disaster in 1994.[5] The focus has been on alarm system performance, alarm reduction techniques, and the display of alarm information in general. It is not the author's intention here to rehash any of the arguments surrounding these issues but to place some of these ideas in the context of control room information.

Whenever operators are dedicated to a particular process in which meaningful alarm information is absent, it is the operators themselves who act as a de facto alarm system. With the modern luxury of many detailed views of the plant, the operator is able to identify abnormal situations as they develop. This is not to say that an operator in an emergency situation would not benefit from meaningful alarm information. On the contrary, this is quite a risky mode of operation because the operator would not to be able to view the entire plant in this manner. In addition, if the operator was required to be absent from the control room for any reason, a colleague would have had to cover and perhaps wind up struggling to manage a process for which he may have little or no training.

In the mode of operation proposed in this chapter, alarm information is very important. As the operator must monitor more processes, he cannot hope to view them all in detail. In emergency or alarm situations, information must be delivered comprehensively to the operator. The previously described method of reporting by exception through operator displays needs to be complemented by an audible alarm facility and a system for managing alarms. Much has been written about the types of facilities that should be available in an alarm management system.[6] In fact, any good DCS or Supervisory Control And Data Acquisition (SCADA) system provides all the recommended functionality in some form.

The process of designing and configuring an alarm system is far more complex than merely selecting the best product. To make an alarm system effective, a great deal of intelligence must be built in to filter, enable and disable, dynamically adjust settings, and otherwise manipulate individual points. This is exacerbated in batch processes due to the many and varied process states.

The key to a good alarm system design is to first focus on designing the alarms themselves. It is a good idea to ruthlessly remove any alarm that does not conform to the definition listed previously. For each alarm there should be a written procedure that details the action that should be taken. There are those in the industry who propose that there should be an informative alarm that requires no action in and of itself but alerts the operator to a developing problem that may soon require action. This level of alarm certainly makes sense, but it is better to get the main system working first and gain confidence in it as this then provides a good foundation for further refinements.

Traditionally, the activation of a safety-related hard-wired trip has been brought to the attention of the operator by use of hard-wired annunciator panels that are separate from the DCS. Although it is often the case that this category of alarm does not conform to the definition given previously, it is still necessary for the operator to be aware that a trip has occurred. It is desirable for the operator to check that the hard-wired trip has operated correctly. However, there is no real reason to keep the hard-wired annunciator panels, as the data could easily be incorporated into the alarm management system of the DCS, and the panels themselves tend to clutter control rooms. The trip initiators should be connected to the DCS anyway, so that they can be incorporated into interlock logic as a mimic of the hard-wired trip action.

Conclusion

Increased automation in the process industries has resulted in fewer process operators. These changes have resulted in a need to change the way that process plants are operated. In particular, there have been changes in the way that process information needs to be presented to the operator.

The intention here has been to develop a generic methodology for the display of batch processing information in the control room. As such, the methodology is not prescriptive in how this should be achieved, but it is assumed that in a modern plant control system the necessary tools will be available to implement any design.

A philosophy of reporting by exception has been adopted so that control room operators may be quickly alerted to abnormal situations. The approach has advocated the adoption of a layered structure for displays coupled with a well-designed and closely integrated system for alarm handling. Displays are kept as simple as possible throughout the structure and careful consideration is given to the use of color. Where detailed information is required it should be accessed using navigational tools rather than cluttering main graphics.

Above all, the emphasis has been on the delivery of key information to the control room. Data that are of no significance to the operator have no place on displays or in alarms and should be avoided. Information should assist the control room operator in his job, not overwhelm him. In a highly automated plant, the job of a control room operator is to prevent loss by intervening to keep the process under control where the DCS has failed to do so. In this sense, the control room operator is only as good as the information he receives.

References

1. Nimmo, Ian. 2000. Ergonomic design of control centres. Paper presented at the IBC Conference: Supporting Control Room Operations, November 2000.

2. McCulloch, John. 2000. Practical principles in operator interface design. Paper presented at the IBC Conference: Supporting Control Room Operations, November 2000.

3. Mill, Robert C. 1994. *Human factors in process operations*. London: Institution of Chemical Engineers.

4. Honeywell European Users' Group Operator Interface Workshop. Quoted in *ACTT News*, November 1999.

5. Health and Safety Executive (HSE). 1997. *The explosion and fires at Texaco refinery, Milford Haven, 24 July 1994*. London: HSE.

6. Bransby, M. L., and J. Jenkinson. 1998. *The management of alarm systems: Health and safety executive contract research report 166*. London: HSE.

Monitoring Multi-recipe Batch Manufacturing Performance

Presented at the WBF
European Conference,
Mechelen, Belgium,
October 11–13, 2004, by

Julian Morris

julian.morris@ncl.ac.uk
Center for Process Analytics and
Control Technology (CPACT)
School of Chemical Engineering and
Advanced Materials, University of Newcastle,
Newcastle upon Tyne, NE17RU, UK

Elaine Martin

e.b.martin@ncl.ak.uk
Center for Process Analytics and
Control Technology (CPACT)
School of Chemical Engineering &
Advanced Materials, University of Newcastle,
Newcastle upon Tyne, NE17RU, UK

Abstract

Quality and consistency are key factors in determining business success. Manufacturing products that satisfy product quality and consistency specifications result in increased productivity and lower overall manufacturing costs. Approaches to achieving consistent high-quality production and enhanced manufacturing performance include Statistical Process Control (SPC) and Six Sigma, with increasing attention now being paid to Multivariate Statistical Process Control (MSPC) methodologies or perhaps better termed "Process Performance Monitoring." In today's process manufacturing environment, a number of issues challenge the application of MSPC-based process performance monitoring technologies. For example, most applications of MSPC have tended to focus on the manufacturing of a single product (e.g., one grade, one recipe) with separate models being developed to monitor individual product types. However, with process manufacturing trends being influenced by customer demands and the drive for product diversification, there has been an increase in flexible manufacturing. Thus in many companies now producing a wide variety of products, there is a real need for process models that allow a range of products, grades, or recipes to be monitored using a single process representation. Three industrial case studies are presented to demonstrate the application of the multi-group performance monitoring approaches.

Introduction

Over the last decade the emphasis in process manufacturing has changed. Quality and product consistency have become major consumer decision factors and are key elements in determining business success, growth, and competitive advantage. Manufacturing products that meet their quality specifications the first time they are produced will result in higher productivity and reduced manufacturing costs through less rework, give-away, and waste. This all contributes to reducing the impact of the process on the environment by minimizing raw materials and energy usage. The achievement of "right first time" production requires a reduction in process variability. Thus the monitoring of process behavior over time to ensure that the key process and product variables remain close to their desired (target) value is essential. This has led to a significant increase in the industrial application of statistical methods to analyze and obtain an enhanced understanding of the process and to the implementation of SPC for process monitoring to obtain early warnings about the onset of changes in process behavior.

Multivariate Process Performance Monitoring (MPPM) is a rapidly growing area of interest for process monitoring. MPPM schemes have typically been based on the statistical projection techniques of Principal Component Analysis (PCA) and Projection to Latent Structures (PLS) and their multi-way extensions for batch processes. Reported practical applications of MPPM have focused on the production of a single manufactured product (e.g. one grade or one recipe), with separate models to monitor different types of products (Kosanovich, Dahl, and Piovoso

1996; Kourti, Nomikos, and MacGregor 1995; Rius, Callao, and Rius 1997; Martin, Morris, and Kiparrisides 1999). However, in recent years, process manufacturing has increasingly been driven by market forces and customer needs, resulting in the necessity for flexible manufacturing to meet the requirements of changing markets and product diversification. Thus with many companies now producing a wide variety of products, there is a real need for process-monitoring models that allow a range of products, grades, or recipes to be monitored using a single process representation.

The elimination of intergroup variation is a prerequisite for statistical process monitoring, so that analysts can focus on intraprocess (intraproduct) variability. This normally requires the construction of separate control charts for each type of product or grade being monitored. In many process-monitoring situations this may be impractical because of the large number of control charts required to monitor the various products and the limited amount of data from which to develop a process representation. An extension to Multi-way PCA (MPCA) and Multi-way PLS (MPLS) that allows the construction of a multiple group model is proposed in this chapter, based on combining the variance-covariance matrices of each of the individual groups. The loadings for the latent variables are then calculated from the pooled variance-covariance matrix of the individual groups. Both MPCA and MPLS approaches have been proposed (Lane et al. 2001; Lane, Martin, and Morris 2003; Martin and Morris 2003).

PLS

A brief overview of the PLS algorithm is presented in this section. A more detailed discussion of the methodology can be found in Garthwaite (1994). The objective of PLS is to determine a set of latent variable scores that "best" describes the variation in the process data set (X) that is most influential on the quality data set (Y). Using these latent variables allows us to construct a set of latent variable scores for the process data (e.g., $T = XW$, where T is the matrix of latent variable scores and W is the matrix of the latent variable loadings). A number of different algorithms have been proposed to derive the loadings for the latent variables associated with PLS. One approach is based on extensions to the Non-linear Iterative Partial Least Squares (NIPALS) method, which regresses the columns of X on Y directly. Consequently, it is not feasible to combine a number of different data sets into a single model.

Lindgren, Geladi, and Wold (1993) present a kernel algorithm for determining the latent variables that is based on the eigenvector decomposition of the variance-covariance matrix. By adapting the kernel algorithm, a multiple group model can be constructed by pooling the individual variance-covariance matrices. In this way, the formal statistical basis for the multiple group model, as given by Flury (1987), can be extended. The variance-covariance approach is based on the hypothesis that the first a eigenvectors of each of the individual variance-covariance matrices span the same common subspace. Although the model introduced by Flury (1987) relates to common principal components, the hypothesis is also appropriate for PLS, since the variance-covariance matrices are of interest. Krzanowski (1984) also shows that the common loadings for the latent variables can be extracted from a weighted sum of the individual variance-covariance matrices.

The pooled correlation (variance-covariance) approach is based on the existence of a common eigenvector subspace spanned by the first a eigenvectors of the individual correlation (variance-covariance) matrices. In practice, the pooled correlation (variance-covariance) approach proposed by Krzanowski (1984) provides a pragmatic method for developing the pooled correlation (variance-covariance) approach by comparing subspaces defined by the eigenvectors associated with the largest eigenvalues. This is a major consideration when determining the method to be used for calculating the latent variables for process monitoring, since it is convention to construct process models using the eigenvectors that correspond to the largest eigenvalues. Thus determining the common latent variables from the pooled correlation (variance-covariance) matrix is particularly appropriate for industrial applications.

Industrial Applications

Batch processes differ from continuous processes in that each variable, j, is measured at k time intervals for a total of I batches. Thus the data set is three-dimensional ($j \times k \times I$). Consequentially, interest lies in both the inter- and intrabatch variability. The application of MPCA or MPLS to the three-dimensional data array associated with batch manufacturing is equivalent to performing standard PCA or PLS on a large two-dimensional data matrix formed by unfolding the original three-dimensional array. The unfolding approach adopted in this chapter is that proposed by Kourti and others (1995). This approach allows the variability between batches to be analyzed by summarizing the variability in the data with respect to both variables and their time variations. The data contained in the two-dimensional matrix is mean centered and scaled prior to applying either MPCA or MPLS. By subtracting the mean of each column from the two-dimensional data matrix, the nonlinearities are effectively removed from the data. The following three case studies are used to demonstrate the multi-group and multi-group, multi-block performance monitoring approaches.

Case Study One: Multi-recipe Manufacturing

The industrial process initially selected to demonstrate the detection and diagnostic capabilities of the multi-group and multi-group, multi-block methodologies is a batch production process in a manufacturing sense. The process includes the two main characteristics of batch operation—flexibility and finite duration. However, in a statistical sense the data matrix is only two dimensional with a single, interbatch source of variability. This is a result of the semidiscrete nature of this particular manufacturing process, with the process variables being measured only once during each batch. This process produces a wide range of household products to meet the demands of a rapidly changing and evolving market. It is comprised of a sequence of individual production steps involving the sequential dosing and mixing of a number of raw materials. During each raw material addition (dosing) a number of process measurements are recorded, including operations and dosing times, flow meter measurements and load cell dose weights, and batch temperatures and the temperatures of the hot and cold process water added to the batch.

There are two types of products produced in the plant. Each formulation can be subdivided into a number of different recipes, which in turn can be further subdivided into a number of different varieties. The formulations are manufactured in separate areas of the plant, with each area having two mixers that are used simultaneously. For some recipes there is a premixing stage for certain ingredients (prior to their addition to the main mix) that is carried out simultaneously with the main mixing process. Off-line analysis of several historical data sets shows that there are distinct differences between the mixers (Lane et al. 2001). As a consequence, a separate model is required for each mixer as well as for each variety of recipe—which is not a practical solution. To demonstrate the application of MPLS, the production of four recipes is analyzed in this case study. Each process mixer is considered a separate "recipe," and thus the process model contains four distinct groups as shown in Table 10.1. It can be observed that recipes 3 and 4 have fewer raw materials and process variables than recipes 1 and 2, which provides an interesting and common industrial challenge for MSPC.

Table 10.1. Composition of the multi-recipe data sets					
Recipe	*Mixer*	*Batches*	*Raw materials*	*Process variables*	*Quality variables*
1	1	19	23	120	1
2	2	21	23	120	1
3	1	20	17	90	1
4	2	29	17	90	1

Figure 10.1 shows a bivariate scores plot demonstrating the problems that arise from combining data sets from different recipes, grades, reactors, and so on (Lane et al. 2001), illustrating the inter- and intravariability that can arise from combining the data into a single data set. In Figure 10.1, the separation between the different recipes and mixers is demonstrated on a bivariate scores plot where the separations of the scores into distinct clusters can be observed. The presence of interrecipe or intermixer variability has a detrimental effect on the detection and diagnostic capabilities of the process model, rendering it insensitive to process faults. In contrast, the pooled variance-covariance matrix, which is constructed as a weighted sum of the individual elements of the individual variance-covariance matrices, provides a sensitivity to the detection of process abnormalities, similar to a single-recipe model.

To address recipes containing different numbers of raw materials and process variables, the pooled variance-covariance matrix is constructed from the individual elements of each of the individual variance-covariance matrices:

$$R_p(k,j) = \frac{(n_j-1)\, R_i(k,j)}{\hat{N}-\hat{g}}$$

where $R_p(k,j)$ is the $(k,j)^{\text{th}}$ element of the pooled covariance matrix, $R_i(k,j)$ is the $(k,j)^{\text{th}}$ element of the individual variance-covariance matrix for recipe I, $\hat{g}$ is the number of recipes containing both the ith and j^{th} variables, and $\hat{N}$ is the total number of observations for the groups containing both the ith and j^{th} variables. If either or both of the variables are not present in a recipe, then there is no contribution to the pooled variance-covariance matrix.

Figure 10.2 shows the reference model, constructed using the pooled variance-covariance matrix of the four groups to calculate the latent variable loadings. The pooled variance-covariance approach's capability to handle data sets containing different numbers of variables provides a powerful advantage. However, care needs to be taken when considering recipes containing different numbers of variables because the inclusion or exclusion of variables may affect the covariance structure of the data set. Cross-validation indicated that the first ten latent variables, explaining 60% of the process variation and 70% of the quality variation, were sufficient to monitor the process.

To demonstrate the detection and diagnostic capabilities of the MPLS model, a batch exhibiting an abnormal product temperature profile was projected onto the reference model in Figure 10.3. The bivariate scores plot of latent variables 3 and 4 clearly identified the batch as being abnormal. In this particular example, by examining the loadings, it was identified that latent variable 3 was associated with the product temperature and that latent variable 4 was associated with the temperature of the process water added to the batch. The monitoring of process

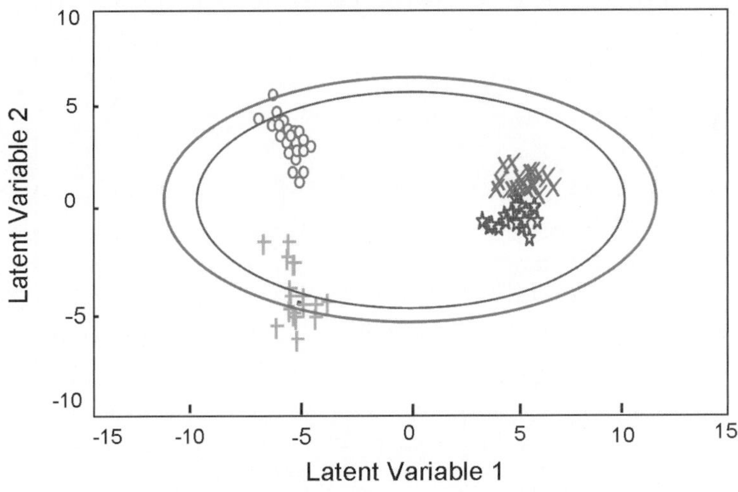

Figure 10.1.   Bivariate scores plot (combined).

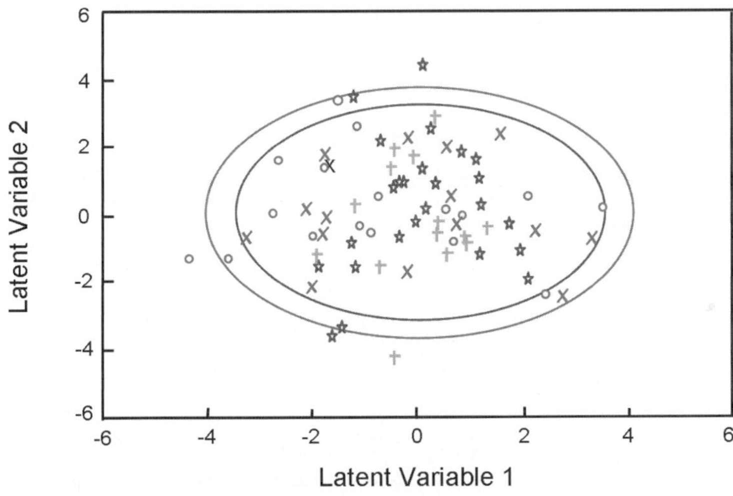

Figure 10.2.   Bivariate scores plot (pooled).

performance takes place on the completion of each raw material dosing step. It can be seen that the observations move away from the center of the in-control region following the first raw material dosing. Subsequent observations continue to drift away from the in-control region and an "out of statistical control" signal occurs after completion of the third raw material dosing.

In the first observation outside the action limits, the contribution plot for latent variable 3 (Fig. 10.4) shows three variables making larger contributions to

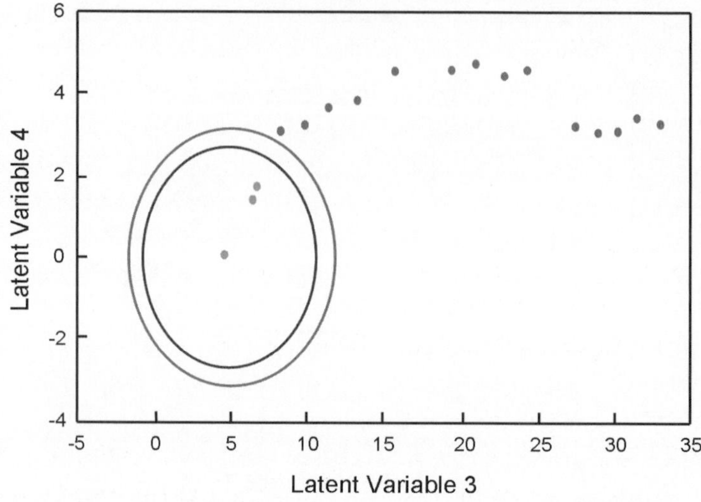

Figure 10.3.   Bivariate scores plot for latent variable 3 versus latent variable 4.

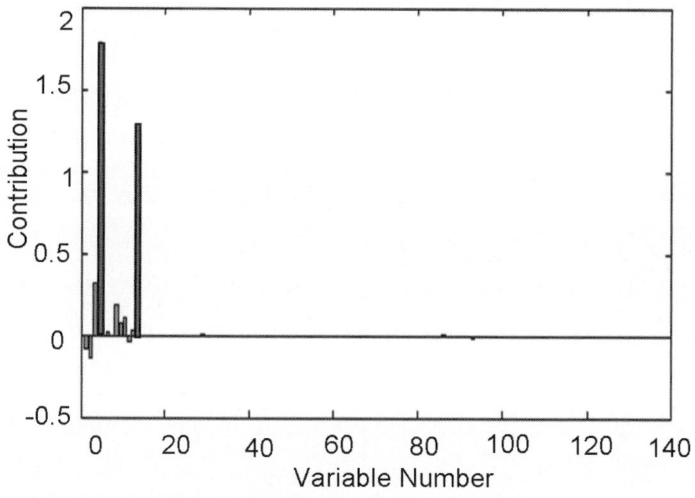

Figure 10.4.   Contribution plot for latent variable 3.

the latent variable scores—variable 4 (cold process water temperature), variable 5 (product temperature following the first raw material dosing), and variable 14 (product temperature following third raw material dosing).

The problem was identified to be a direct result of the failure of the process water cooling system, which caused water to enter the batch above the required temperature and led to an abnormally high product temperature. As the product temperature is monitored following each raw material dosing, the latent variable scores continue to move away from the in-control region. A second dosing of process water also occurred later on in the batch, which can be observed by the small change in the direction of the scores following the seventh raw material dosing (Fig. 10.3). At the time of the application, no remedial action was taken during the production process, and one of the motivations for the on-line application was to allow operators to be more proactive when a processing problem occurs.

Case Study Two: Hierarchical Multi-block Modelling

The objective of hierarchical principal component analysis and Multi-block Partial Least Squares (MBPLS) is to divide the process into a number of meaningful blocks and then apply PCA or PLS to the individual blocks. The individual block scores are combined into a consensus matrix, and PCA or PLS is again applied. The consensus scores are then used to monitor the overall process. Consequently, when a process disturbance occurs, it is possible to isolate the block in which the disturbance has occurred. The individual block scores are then used to provide information about the specific variables that are indicative of the source of the disturbance. A schematic of the MBPLS monitoring scheme is given in Figure 10.5. Here the consensus monitoring chart detects the process disturbance, and the corresponding contribution plot identifies the block in which the disturbance occurred. Finally, the contribution plot for that particular block identifies the variables indicative of the disturbance.

The concept of hierarchical multi-block PCA and PLS was initially introduced around 1986. More recently, papers on hierarchical algorithms have been published by Wold and others (1996) and Westerhuis and others (1998). The original algorithm was termed Consensus PCA (CPCA) and was presented as a method

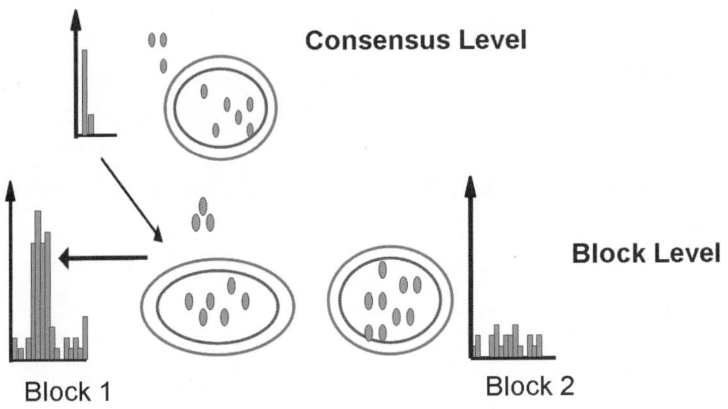

Figure 10.5.   Multi-block process performance monitoring.

for comparing several blocks of descriptor variables measured on the same object. In this algorithm, the data are divided into a number of blocks, and the block loadings are then computed. In turn, these are used to calculate the block scores, which are combined into a "super block." The super block loadings are then computed and normalized to unit length. Finally, a super score is computed. After convergence, all the blocks are deflated using the super scores, and a second super score—orthogonal to the first—is computed using the block residuals.

In the process manufacturing study described previously (case study one), several of the recipes in the manufacturing process undergo a premixing as well as a main mixing stage during manufacturing. This makes multi-block PLS a particularly attractive methodology for building a monitoring model, as there is a natural blocking of the variables. To demonstrate the multiple group, multi-block PLS methodology, three recipes are analyzed in this case study. This results in six data sets, since each recipe can be produced in either of two mixing areas. The composition of the data set is summarized in Table 10.2.

In this case, each recipe contains the same number of raw materials. This is because the recipes are very similar and only differ in terms of coloring. In this application, the individual colorings are treated as being the same raw material because the coloring is only considered a "minor" raw material and perceived to

Table 10.2.   Composition of the process data sets

Recipe	Mixing area	Batches	Quality variables
1	1	48	3
1	2	43	3
2	1	16	3
2	2	27	3
3	1	51	3
3	2	31	3

Premixer		Main mixer	
Raw materials	Process variables	Raw materials	Process variables
10	41	17	84
10	41	17	84
10	41	17	84
10	41	17	84
10	41	17	84
10	41	17	84

have no influence on product quality. As in the previous study, a set of monitoring charts are constructed using data collected when the process was manufacturing acceptable quality product and there were no assignable causes of variation present in the data. Three sets of monitoring charts are required—one for each individual block and one for the overall process (i.e., the consensus model).

The fault detection and diagnostic capabilities of the multi-group, multi-block PLS model are demonstrated using a batch where a raw material overdosing took place. As with the previous example, each latent variable score "x" represents the status of the current "on-line" batch following each successive raw material dosing. The consensus scores plot for monitoring the overall process (Fig. 10.6) shows an "out of statistical control" signal at the sample point following the overdosing. Unlike the previous example, the latent variable scores cluster in the center of the in-control region, indicating "good" operation until the overdosing occurs. Following the overdosing, there is an abrupt departure from the in-control region.

Examination of the contribution plot for latent variable 2 (Fig. 10.7) identified the premixer as the process subsection where the overdosing occurred. Inspection of the

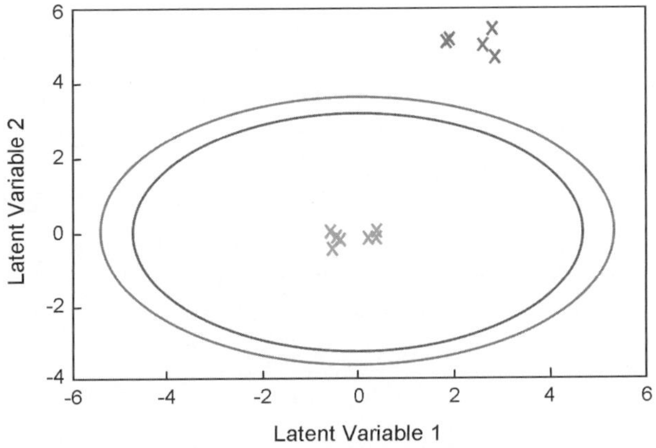

Figure 10.6.   Latent variable scores plot (consensus chart).

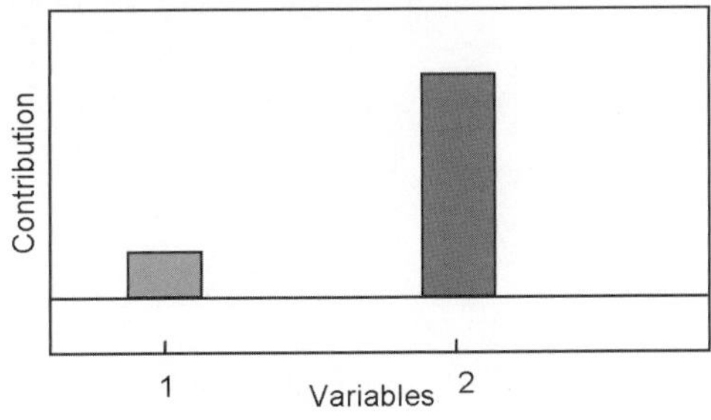

Figure 10.7.   Contribution plot for latent variable 2 (consensus chart).

monitoring chart for the premixer (Fig. 10.8) also identified the "out of statistical control" signal at the sample time following the overdosing. In contrast, no out-of-control signal was identified from the latent variable scores plot of the main mixer (Fig. 10.9).

For this particular fault, both the consensus and the premixing monitoring charts detect the fault at the same time. This is due to the small number of variables included in each block and the abrupt impact that the fault had on the latent variable scores. This is a direct consequence of the individual block monitoring charts only monitoring the deviation of the variables contained within that particular block and not the whole process, which is the case with the consensus monitoring charts.

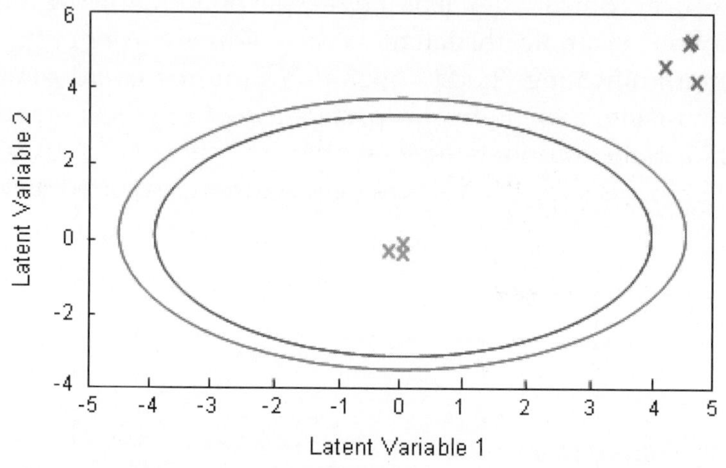

Figure 10.8.　Latent variable scores plot (premixer chart).

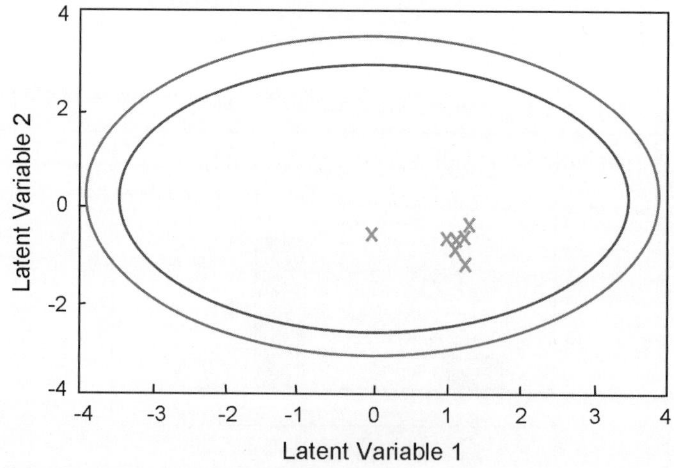

Figure 10.9.　Latent variable scores plot (main mixer chart).

Investigation of the contribution plot for latent variable 2 (Fig. 10.10) identified variables 9 (dosing time), 10 (load cell dose weight), 11 (flow meter dose weight), and 12 (product temperature) as being responsible for the "out of statistical control" signal. From this information, it was concluded than an overdosing of raw material 3 had occurred. A process engineer later confirmed that the overdosing was caused by a malfunction of a dosing valve that controls the amount of raw material being added into the batch. Prior to the application of MSPC, a similar problem might have caused the manufacturing of a number of "out of specification" batches before the problem was identified by chance.

In such manufacturing situations only a few "key" variables may be monitored, and faults impacting the remaining variables can be missed. The study has demonstrated that such processes could be monitored effectively using multi-group, multi-block MSPC. Contrary to the premixing monitoring chart, the latent variable scores for the main mixer remain clustered in the center of the in-control region throughout the manufacturing of the entire batch (Fig. 10.9). This is because the fault detected in the premixer has no impact on the variables being monitored in the main mixer. However, as the product from the premixer is added to the main mixer, the product in the main mixer will be affected by the premixer fault unless the process is halted when the fault occurs.

Case Study Three: Application of Multi-group Modeling to Microelectronics Manufacturing

To demonstrate the wider capabilities of MPCA performance monitoring, three sets of data from a metal etcher process are considered (Wise et al. 1999). Data are supplied from an A1-stack etching process that is performed using a Lam 9600 plasma-etching tool. The objective of the process is to etch the NiN/A1-0.5% Cu/TiN/oxide

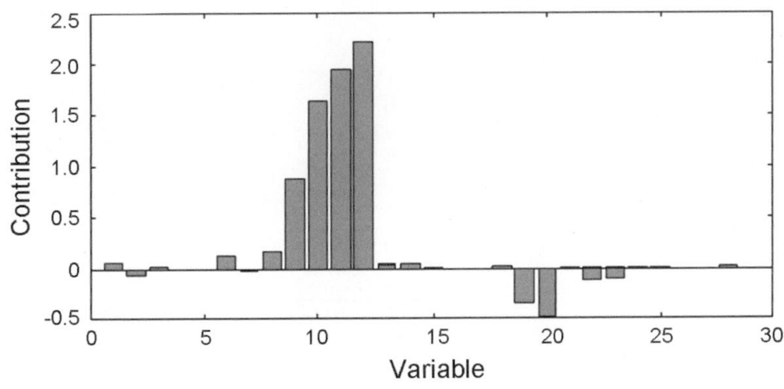

Figure 10.10.   Contribution plot for latent variable 2 (premixer).

stack with an inductively coupled BCI3/CI2 plasma. The standard manufacturing process consists of a series of six steps. The first two allow for the achievement of gas flow and stabilization. Steps three and four consist of a brief plasma ignition and the main etch of the A1 layer terminating at the A1 endpoint, respectively. Step five provides an over-etching of the underlying TiN and oxide layers, while step six is associated with the venting of the chamber.

The etching of an individual wafer is analogous to a single batch in a chemical process. Changes in the process mean are the result of a residue building up on the inside of the chamber following the cleaning cycle, differences in the incoming materials resulting from changes in the upstream process, and drift in the process-monitoring sensors themselves. As a result of the changes in the process mean, three distinct operating levels can be identified in the data set (Fig. 10.11). When the data are combined into a single data set, the scores of Principal Component 1 (PC1) and Principal Component 2 (PC2), which are not shown but similar to the latent variables plot shown in Figure 10.1, reflect these discrete operating levels.

In this situation, the major source of variation explained by the individual principal components is the variation of each variable from the overall mean of the data set. Hence this identifies the different operating conditions of each variable. This between group variation present in the data set causes the principal component scores to cluster according to which operating region they represent. When such clustering occurs, there are two issues that impact process performance monitoring: (1) the control limits may be conservative and assignable cause process events may not be detected and (2) an assignable cause reflected in the movement of a principal component score into another cluster when the operating conditions have not been changed may result in the real process event not being detected. As a consequence of the changing mean levels observed, the process data were

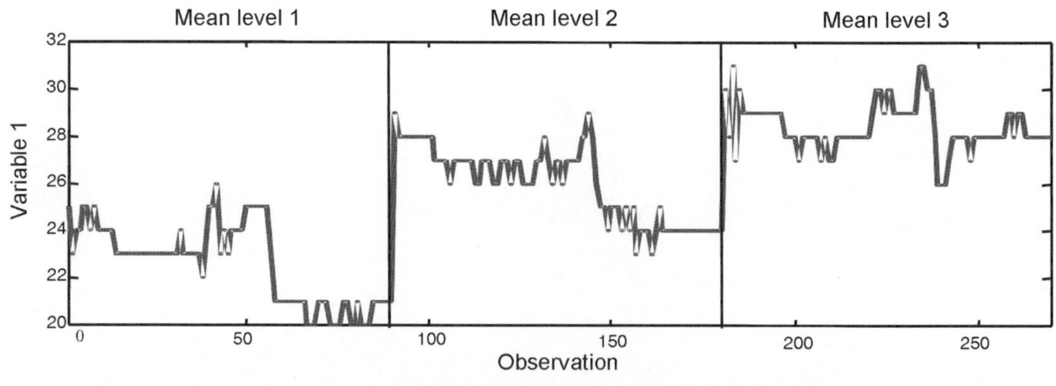

Figure 10.11.   Time series plot of mean levels resulting in distinct clusters in the PC1 versus PC2 score plot.

divided into three subsets—one for each operating level. The composition of each of these data sets is presented in Table 10.3.

A pooled covariance reference model for the multiple group application was constructed using the three data sets. By analyzing each of the data sets, it can be inferred that the different operating levels share common characteristics that determine the process behavior, and as a consequence the use of the multiple group modeling approach is validated. Ten principal components were selected from cross-validation explaining 68%. A bivariate scores plot for PC1 and PC2 of the pooled covariance model is shown in Figure 10.12, indicating that the scores are independent and identically distributed, thus indicating that the multiple group monitoring model provides a good representation of the overall etch process.

To evaluate the detection and diagnostic capabilities of the multiple group model, a data set containing an increase in the Transmission Control Protocol (TCP) power is projected onto the reference model. This is done in a manner so as to simulate an on-line monitoring situation. Each observation that is projected onto the monitoring chart represents the status of an on-line batch at successive sampling points during the etch process. The bivariate scores plot of PC1 and PC2 (Fig. 10.13) detects the change in the operating conditions as a slow drift away from the center of the in-control region. At the beginning of the etch run, the principal component scores lie in the center of the in-control region (x). After the first few sample points, the scores gradually drift away from the center toward the control limits. An "out of statistical control" signal is flagged as the scores cross the action limits. In this particular example, no remedial action is taken and the scores continue to drift away from the in-control region until the conclusion of the process run. Analysis of the contribution plot (not shown) identifies subtle changes in two pressure measurements as can be observed in Figure 10.14.

Conclusion

The industrial applications described in this chapter have demonstrated the extension of standard MSPC methodologies to processes where different products or

Table 10.3. The metal etcher data sets			
Operating level	*Observations*	*Variables*	*Batches*
1	90	17	17
2	90	17	16
3	90	17	16

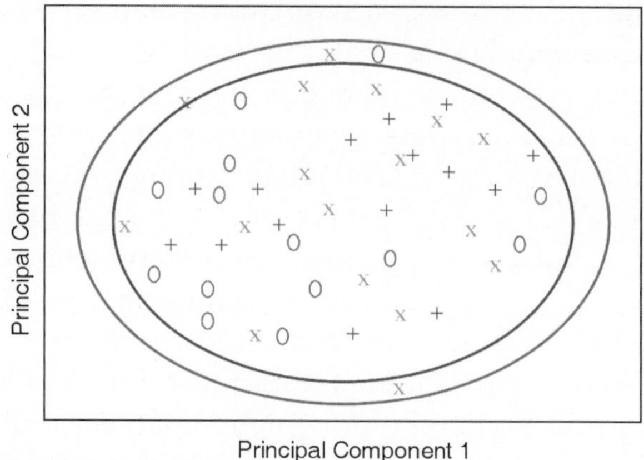

Figure 10.12.   Bivariate scores plot (pooled).

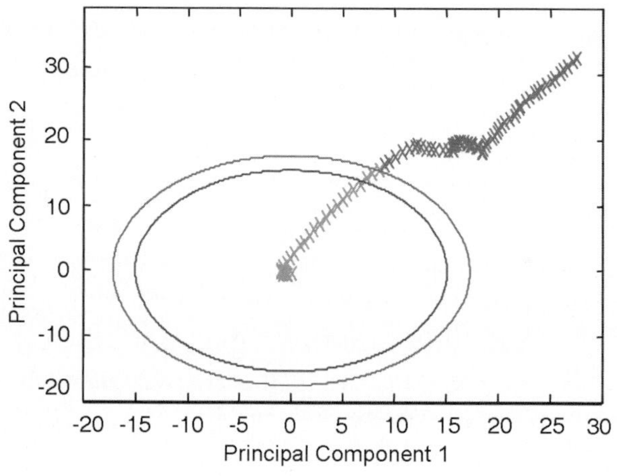

Figure 10.13.   Bivariate scores monitoring plot.

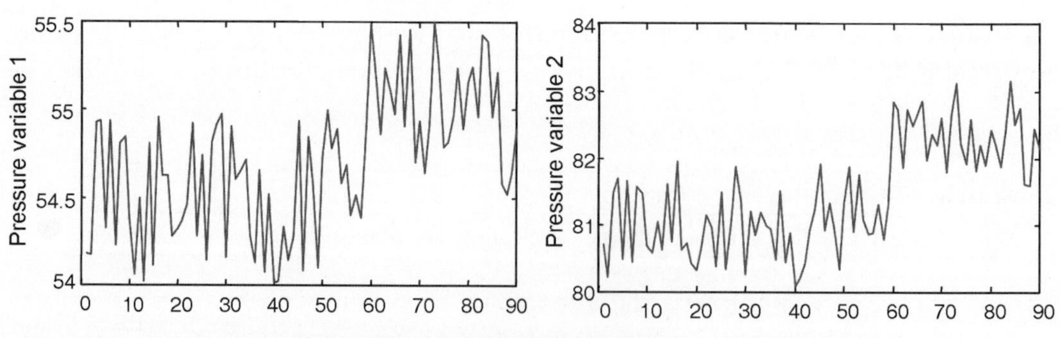

Figure 10.14.   Time series plots of two pressure variables.

recipes are produced in different vessels with different numbers of variables being monitored. In the manufacturing application, the eigenvectors of the pooled variance-covariance matrices give a good representation of the production process, even though in some cases the recipes contain a different number of raw materials and consequently the data sets contain different numbers of variables. The results show that both multi-group and multi-group, multi-block monitoring models have good detection and diagnostic properties. The latent variable scores plots detect batches that are out of statistical control, and the corresponding scores contribution plots identify the variables responsible for the out-of-control signal. Contribution plots give explicit information about process abnormalities.

The applications presented in this chapter are from manufacturing processes where the amount of data from each distinct set of operating conditions or product grade is limited. Prior analysis of these individual data sets suggests that the data available are not sufficient to give a true reflection of the underlying process variability. By applying the pooled correlation (variance-covariance) approach, all the products being manufactured could be monitored using a small number of monitoring charts. In this way, the cost and time required to update the models can be significantly reduced, allowing the faster on-line implementation of process performance monitoring schemes.

Acknowledgments

The authors acknowledge the EU ESPRIT Project "Process Diagnostics for Plant Performance Enhancement" (PROGNOSIS No. 22281), Dr. Steven Lane, and the contributions to the on-line industrial multi-recipe application by Dr. Chris Hawkins (MDC Technology), Emerson Process Management.

References

Flury, B. N. 1987. Two generalizations of the common principal component model. *Biometrika* 74: 59–69.

Garthwaite, P. H. 1994. An interpretation of partial least squares. *Journal of the American Statistical Association* 89: 122–27.

Kosanovich, K. A., S. Dahl, and M. J. Piovoso. 1996. Improved process understanding using multi-way principal component analysis. *Industrial and Engineering Chemistry Research* 35: 138–46.

Kourti, T., P. Nomikos, and J. F. MacGregor. 1995. Analysis, monitoring and fault diagnosis of batch processes using multiblock and multiway PLS. *Journal of Process Control* 5: 277–84.

Krzanowski, W. J. 1984. Principal component analysis in the presence of group structure. *Applied Statistics* 33(2): 164–68.

Lane, S., E. B. Martin, and A. J. Morris. 2003. Batch monitoring through common subspace models. Proceedings of the IFAC Conference ADCHEM, Hong Kong.

Lane, S., E. B. Martin, A. J. Morris, and R. A. G. Kooijmans. 2001. Performance monitoring of a multi-product semi-batch processes. *Journal of Process Control* 11: 1–11.

Lindgren, F., P. Geladi, and S. Wold. 1993. The kernel algorithm for PLS. *Journal of Chemometrics* 7: 45–59.

Martin, E. B., and A. J. Morris. 2003. Monitoring performance in flexible process manufacturing. Proceedings of the IFAC Conference ADCHEM, Hong Kong.

Martin, E. B., A. J. Morris, and C. Kiparrisides. 1999. Manufacturing performance enhancement through multivariate statistical process control. *Annual Reviews in Control* 23: 35–44.

Rius, A., M. P. Callao, and F. X. Rius. 1997. Multivariate statistical process control applied to sulfate determination by sequential injection analysis. *The Analyst* 122: 737–41.

Wise, B. M., N. B. Gallagher, S. Watts Butler, D. D. White Jr., and G. G. Barna. 1999. A comparison of principal component analysis, multiway principal component analysis, trilinear decomposition and parallel factor analysis for fault detection in a semiconductor etch process. *Journal of Chemometrics* 13: 379–96.

Transfer Lines as Units in an ISA-88 Framework

Presented at the
WBF North American
Conference, Chicago, IL,
May 16–19, 2004, by

Todd A. Brun
Senior Project Engineer
tabrun@ra.rockwell.com
Rockwell Automation, 15458-B N. 28th Avenue,
Phoenix, AZ 85053, USA

Abstract

Traditional process models typically view a transfer line between units either as a physical extension of the batching vessel or as a shared Equipment Module (EM). The valves in the transfer line are then convenient places to establish the boundary of the upstream or downstream unit.

A "transfer panel" (Fig. 11.1)—also known as a "hook-up station" or a "flow-verter"—is a collection of piping ports that use U-bend pipes or flexible hoses to connect one port to another and is frequently used in food and pharmaceutical processes to facilitate different operations. Transfer panels also offer greater flexibility in equipment configurations and also provide physical isolation between processes.

While transfer panels provide these benefits, they also challenge traditional methods of applying ISA-88.01. This chapter discusses how the introduction of transfer panels into a piping design elevates the transfer line to the status of an ISA-88.01 unit and illustrates how this view simplifies the design of the associated control system.

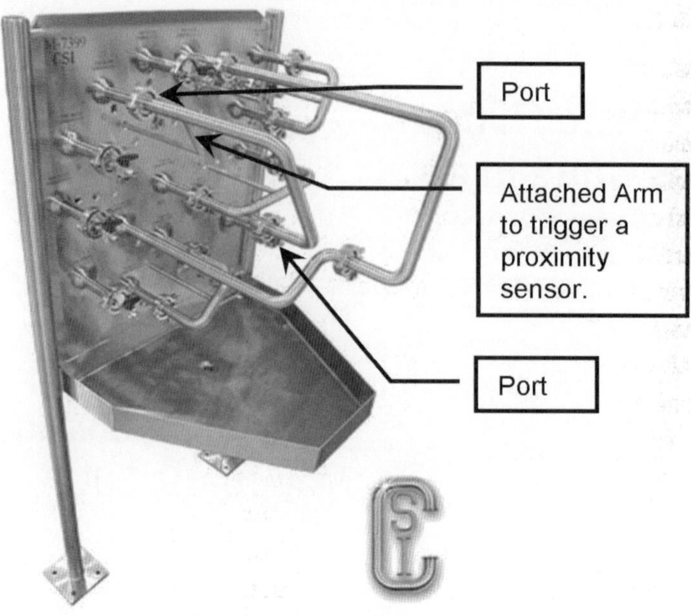

Figure 11.1.   Example of a transfer panel. Note the ports and U-bends.
Picture courtesy of Central States Industrial, 2700 Partnership Blvd., Springfield, MO 65803.

The Traditional View of Transfer Lines

Consider a simple system of one upstream tank, an outlet valve, and a downstream tank (Fig. 11.2). If the two vessels have processes unto themselves, then each is a separate unit as defined by ISA-88.01. The most difficult part of the design is establishing where the boundary between them is in the continuum between the first tank and the second tank. Boundaries imply ownership. Once the designer establishes which tank owns the valve, then the unit boundaries are immediately evident. In this arrangement, there is no real "transfer line." The pipes that connect the tank to the valve are merely extensions of the tank.

Now consider the system shown in Figure 11.3, consisting of one upstream tank, one upstream "inlet" valve, one downstream "outlet" valve, and one downstream tank. It would be normal to reach a preliminary conclusion that each valve belongs to the tank closest to it. However, ownership of the intermediate pipe between the valves needs to be defined.

In the food and pharmaceutical industries, the issue of the intermediate pipe in Figure 11.3 presents serious problems, especially in terms of sanitization. This is where the transfer panel enters the picture.

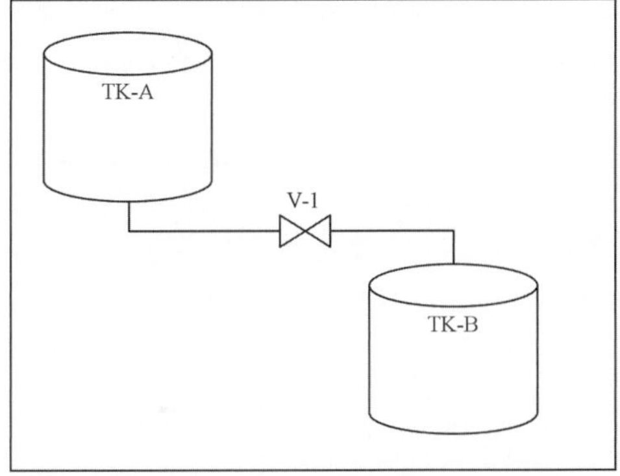

Figure 11.2.   Upstream and downstream tank linked with a transfer line and single valve.

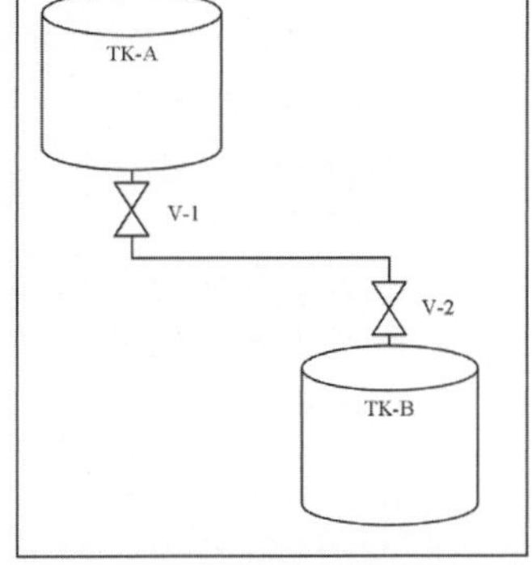

Figure 11.3.   Upstream and downstream tank linked with a transfer line and two valves.

A Brief Digression about "Ownership"

Ownership of "things" is one of the most fundamental concepts to comprehend when designing effective automated control systems. Nebulous delineation of ownership yields chattering of outputs, corruption of data, and uncertain performance of control strategies. But this should come as no surprise. Questions of ownership have been a major source of conflict plaguing mankind since the beginning, and its impact on automated control systems is no different.

Unit Boundaries and the Transfer Panel: Part 1

The transfer panel at each end of the intermediate pipe (Fig. 11.4) creates a physical break in the line so that the line can be easily re-piped to become part of a variety of different flow paths. This break allows for cleaning and sanitization of the line separately from the upstream or downstream tanks. As shown by Figure 11.1, a transfer panel consists of a piece of sheet metal that supports the ports to which the U-bends are connected. Figure 11.1 also shows that transfer panels and their U-bends can be quite elaborate.

The upstream transfer panel (labeled "XP-A" in Fig. 11.4) allows the tank A outlet (port A) to be configured to a steam trap (port D), a Clean In Place (CIP) return header (port C), or a transfer line (port E). Likewise, the transfer line can be configured to a steam supply (port F), the CIP supply header (port B), or the outlet of tank A (port A).

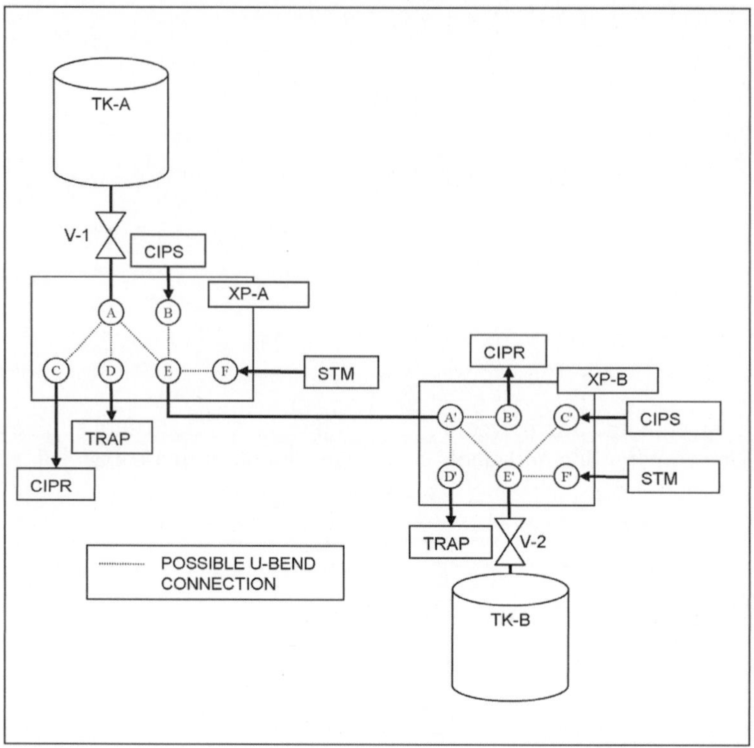

Figure 11.4.   Transfer line with upstream and downstream transfer panels.

Focusing now on the downstream transfer panel (labeled "XP-B"), the inlet of tank B (port E′) can be configured to either a steam supply (port F′), the CIP supply header (port C′), or the outlet of the transfer line (Port A′).

These two transfer panels present us with some interesting combinations of connections: When the outlet of tank A is connected to the transfer line (XP-A ports A through E) and the transfer line is connected to the inlet of tank B (XP-B ports A′ through E′), we get a configuration that is identical to Figure 11.3. The transfer line can be connected to the steam supply on the upstream side (XP-A ports F through E) and to the steam trap on the downstream side (XP-B ports A′ through D′). This frees up the outlet of tank A and the inlet of tank B for other configurations. When the outlet of tank A is connected to the trap (ports A through D), it is not possible for the transfer line to be connected to the tank because port A is already occupied.

This last fact leads us to the possibility of doing one type of operation in tank A (such as Sterilize In Place [SIP]), while doing an entirely different operation in the transfer line (such as CIP). Different operations done simultaneously? Remember the ISA-88.01 rule: operations are done on units. SIP of the tank makes sense, but what do we make of simultaneous CIP of the transfer line?

What Distinguishes a Tank from a Line?

This ought to be a simple question. Our first encounters with a tank and a transfer line can lead to the following *overly simplistic* definitions:

- A tank is a big thing that holds liquid for processing.

- A transfer line is a long, skinny connector that transports liquid from one tank to another.

Unfortunately, our first encounters with tanks and lines can likely leave us with lasting, simplistic impressions such as these that can form the basis of paradigms that can cloud our vision. Table 11.1 shows a list of characteristics that are often associated with tanks and how these characteristics are manifested in transfer lines. This list is certainly not all inclusive, but the point is that the distinction between tanks and transfer lines is not as clear-cut as first impressions would lead us to believe. The more you add to the list, the more it seems that the only thing that distinguishes a tank from a line is its aspect ratio, and even that stereotype is a candidate for challenge!

Now back to the issue of unit boundaries and the transfer panel.

Unit Boundaries and the Transfer Panel: Part 2

In Figure 11.4, we have an upstream tank unit, a downstream tank unit, a transfer line, and two transfer panels separating the transfer line from the tanks. From the operational standpoint, each tank is its own unit with its own distinct set of unique operations such as CIP, SIP, and batching.

Table 11.1. Comparison of tanks and lines	
Typical characteristics of a tank unit	Equivalent characteristics of a transfer line
Holds a single batch	Petroleum transmission lines are hundreds of miles long and are capable of holding "a batch" with plenty of room to spare.
Temperature control of the tank's contents	Jacketed pipes are used for the heating and cooling of its contents.
Agitation to disperse ingredients in the tank	In-line agitation is common in ice cream factories as a means of dispersing chunky ingredients to help disperse them into the ice cream mix flowing to the filling machine (think cookies 'n cream, walnuts, or black cherries).
Pressure control within the tank	Backpressure valves are used to maintain pressure within the transfer line (often a requirement in sterile lines).

But what about the transfer line? It, too, has its own set of unique, stand-alone operations of CIP and SIP. This is the key point: We run recipes on a tank that affects the material in the tank and the state of the tank. Likewise, we run recipes on the transfer line that affect both the contents (unsterile air is a type of content, just like sterile product) and the state of the transfer line. Previously, we saw how a transfer line can look like a tank. Now we just learned that a transfer line can act like a tank—at least in having separate operations.

Hmm . . . If it looks like a tank (or more precisely, a unit) and if it acts like a unit, could it be that it actually is a unit?

If It Looks Like a Unit and Acts Like a Unit, Then Treat It as a Unit

The author's first real encounter with this point of view came while designing a batching system consisting of four batching tanks and two transfer lines that led to two separate fillers (Fig. 11.5). The transfer lines were equipped with transfer panels at each end that allowed each line to be cleaned and used in production independently of the other.

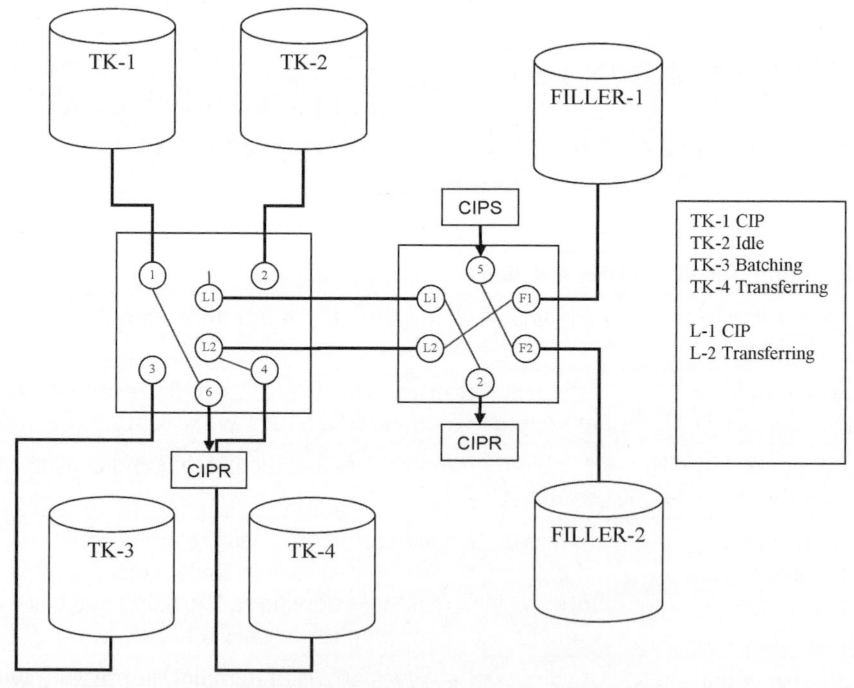

Figure 11.5.   Actual project.

It was natural to think of each batching tank as its own unit with its own control screens and operational unit status displayed on the Human Machine Interface (HMI) control screens. The difficulty arose when the author attempted to incorporate the transfer line into both the documentation of one of the batching units as well as the Programmable Logic Controller (PLC). Every attempt to subordinate the transfer line to a tank unit led to difficulties and inconsistencies.

The fact that it was just as important to display an independent status for each transfer line as it was to display independent status for a batch tank led the author to question—if it looks like a tank, does it act like a tank? The affirmative answer led immediately to the conclusion that if it looks like a unit and acts like a unit, then treat it like a unit. As a result, each transfer line received its own section in the requirements document, its own section in the design documents, its own section of code and data structures in the PLC, and its own unit status shown on the HMI, as well as its own operational control screens.

What was first a complicated design became very flexible, modular, and repeatable: a "copy and paste" with simple search-and-replace edits quickly created the second transfer line in documentation, code, and screens.

Conclusion

Transfer panels allow for the flexible configuration of piping systems. They also allow for concurrent operations to be executed in the transfer line independent of the upstream or downstream units. It is this independent and concurrent nature of the operations that elevate the transfer line to the full-blown status of an ISA-88.01 unit. Treating the transfer line as an equal to a batching tank may seem unnatural, but it will lead to simpler designs that will make your life as a designer easier.

Editor's Note

The need to design a transfer line as an ISA-88.01 Unit arises from the particular control system used and the need to have separate displays. Other systems may allow displays for ISA-88.01 EMs.

Applying ISA-88.01 to Polyethylene Production

Presented at the WBF
North American Conference,
Philadelphia, PA,
March 24–26, 2008, by

Chris Morse
Product Manager
chris.morse@honeywell.com
Honeywell Inc., Lovelace Road,
Bracknell, RG12 8WD, UK

Abstract

The production of polyethylene is considered a classic continuous process. However, a number of the manufacturing challenges associated with polyethylene production are also common to batch manufacturing. These common factors include raw materials management, traceability, and production control. After a detailed analysis of the typical automation requirements, many principles of ISA-88.01 can be applied to the majority of the polyethylene production process if special considerations are taken into account.

The chapter analyzes a newly constructed polyethylene plant where ISA-88.01 was applied to the automation system to ensure flexibility and future supportability and to provide certain business benefits. The overall solution, challenges, and differences from a classic batch process perspective are considered.

Introduction

The nature of batch processes makes them inherently more complex to automate than continuous processes in plants of a similar scale. Since its release, ISA-88.01 has greatly improved the automation of batch processes by providing a standard structure and standard naming conventions. Vendors have been able to standardize products, and best practices are easily shared by all in the industry.

Most continuous processes include sequential elements for switching duty, grade changes, startup, and shutdown. These are executed relatively infrequently compared with the duration of cycle time in batch processes and are often executed manually. This can result in inconsistent execution, operator errors, and inefficiencies that sometimes can generate health, safety, or environmental incidents.

It is the author's experience, having worked in both batch and continuous environments, that applying the principles of ISA-88.01 to sequential aspects of continuous processes can benefit engineering efficiency, operability, and lifetime supportability. This chapter is based on a real project executed in Europe.

Background

At an early phase of the project, the end user engineering team took the opportunity to ensure the business performance of the plant by selecting automation technology that would give the operating team the following:

1. A clear view of planned and "in-progress" production, viewable by all parties

2. Traceability of raw materials through to finished products

3. Tools to convert "data" into information to support continuous improvement

4. Minimal customization to ensure the ease of future support

5. Automation of transitions between products where practical

6. A high level of system availability to ensure continuity of production

Although polyethylene production is a continuous process, it does have many sequential or batch elements. Selecting technology based on ISA-88.01 helped achieve all these objectives—particularly item 4.

Batch and Continuous Polymerization

In a sample batch polymerization reaction using class-based concepts (Fig. 12.1), materials are charged in discrete additions with defined movements between vessels (e.g., V10 > R20 > B60).

Raw materials' tracking is linked to the final batch (e.g., 1101). The boundaries between batches 1100 and 1101 are easily identified by an event such as the completion of a procedural element or a material transfer. A typical batch cycle time would be 12 hours and a batch size equal to 10 tonnes. The level of automation is high, and limited manual interaction would be expected, typically including sampling and adding raw materials in small quantities.

The first noticeable difference in a continuous polymerization process is the cycle time and size of a batch. A batch may represent 3 to 5 days of production and 20,000 tonnes. In terms of raw materials, it may be made up of (a) continuous raw material feeds (in the case of the referenced project ethylene), (b) multiple batches of material that are specific to the product produced and

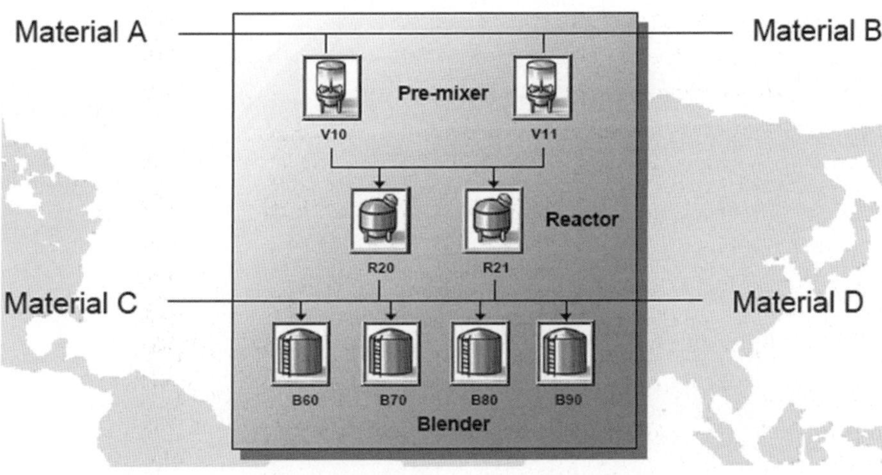

Figure 12.1.   Batch polymerization process.

Equip	V10		R20	B80				V11		R21	B90		
Mats	A	B	Transfer	Transfer	C	D		A	B	Transfer	Transfer	C	D
Batch	1100							1101						

Figure 12.2.   Chart for batch polymerization.

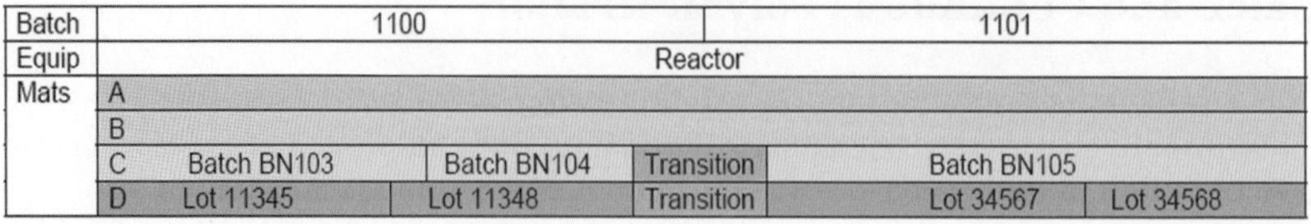

Batch	1100			1101	
Equip	Reactor				
Mats A					
B					
C	Batch BN103	Batch BN104	Transition	Batch BN105	
D	Lot 11345	Lot 11348	Transition	Lot 34567	Lot 34568

Figure 12.3.   Chart for continuous polymerization.

mixed during production for continuous injection into the reactor, and (c) multiple raw material "lots" injected continuously into the reactor.

The second significant difference is that there is no specific event that defines the end of a batch, as the reactor and extruder cannot be stopped and restarted without an unacceptable quantity of off-spec material being produced. Instead the transition between batches involves a period of changing process parameters and raw materials, normally with a period of off-spec production (the quantity of off-spec material being less than that for a complete shutdown and startup). The level of automation in such plants beyond interconnected basic control modules is traditionally low. The transitions are normally executed manually. Timely and consistent execution of these transitions reduces costs and improves quality.

Upon leaving the reactor, the polymerized material is extruded and cut into pellets. It is stored initially in a 500-tonne degassing silo from which residual ethylene is recycled. Material is then transferred to a 500-tonne storage silo before shipment via bulk container. For the purposes of traceability, a number of interrelated "batches" therefore exist:

1. Reactor batch (production order)

2. Material C batch

3. Material D lot

4. Degassing and storage silo batch (both 500 tonnes)

5. Container batch

Applying ISA-88.01 to Achieve Our Objectives

The project engineering team could see that ISA-88.01 could be applied to several aspects of the process and that this would enable the utilization of standard batch software, establish concepts, and help meet our six objectives (as listed previously).

1. A Clear View of Planned and In-progress Production Viewable by All Parties

This is accomplished via DCS displays. All production stakeholders have the same view of the current and planned production orders and other batch types. DCS batch functions for Procedures, Unit Procedures, and Phases are used in the reactor and mixing areas. Recipe parameters for different products are managed at the controller level.

2. Traceability of Raw Materials through to Finished Products

A batch reporting package is used to trace raw materials and relate batches to process conditions. The continuous nature of the process makes this complex. For a particular silo batch, the start and end time of the fill are known. To calculate which batch of raw material C was used, it is necessary to apply a timed offset for the delay between the reactor and the silo. This offset is not constant and depends on the production rate, which varies with reactor and extruder conditions. Techniques from the continuous process industry and additional instrumentation have been employed to calculate this offset as accurately as possible. The smallest level of granularity possible for traceability is one silo.

3. Tools to Convert "Data" into Information to Support Continuous Improvement

Standard batch reporting functions such as traceability and batch genealogy (discussed later on in this chapter) are used to present batch relationships in a quick and user-friendly manner.

4. Minimal Customization to Ensure the Ease of Future Support

The end user had previously experienced high life-cycle support and migration costs for similar projects due to the degree of customization. Wherever possible, standard batch software products are used to protect the end user's investment by reducing the long-term cost of support for upgrades to the DCS, process changes, and migration to new Microsoft operating systems. Standard Human Machine Interface (HMI) objects ergonomically developed for batch displays were also utilized, thus reducing the level of engineering-intensive display development.

5. Automation of Transitions Where Practical

The decision to automate transitions between production orders helps ensure consistency of execution and timely execution for each step and minimizes the quantity of "off-spec" products produced. Functions in the transitions (e.g., "ramp pressure from x bar to y bar over 20 minutes") use similar batch phase logic, so ISA-88.01 phases are used to execute these controller functions.

6. A High Level of Availability to Ensure Continuity of Production

The potential nonavailability of a server-level application when a new production order is scheduled to start is seen as highly disruptive to production. The use of standard batch technology means that all production order execution takes place at the controller level. No server-level applications are required to produce a product or to change to the next production order. In the ISA-88.01 activity model, Recipe Management, Process Management, Unit Supervision, and Process Control are all executed in a redundant controller.

ISA-88.01 Structure

The structure used in the plant is fairly simple in batch terms, since there are only two units that need to be defined: Reactor (which includes the extruder) and Mixer. The procedural differences between product grades were sufficiently small, so only a single set of procedures, unit procedures, and phases were configured. The number and complexity of procedural elements were sufficiently small and consistent across products, so the operation level of the procedure model was omitted. The structure depicted in Figure 12.4 is the result of these decisions.

Beyond the reactor, ISA-88.01 has not been applied. No processing of any type is carried out in the degassing or storage silos, which ISA-88.01 would typically define as units. With the large number of these silos and the simplicity of the material transfers after this point, applying ISA-88.01 would have added too much complexity to the system without realizing added benefits. Instead, a simple batch record was created recording material movements and certain attributes for traceability.

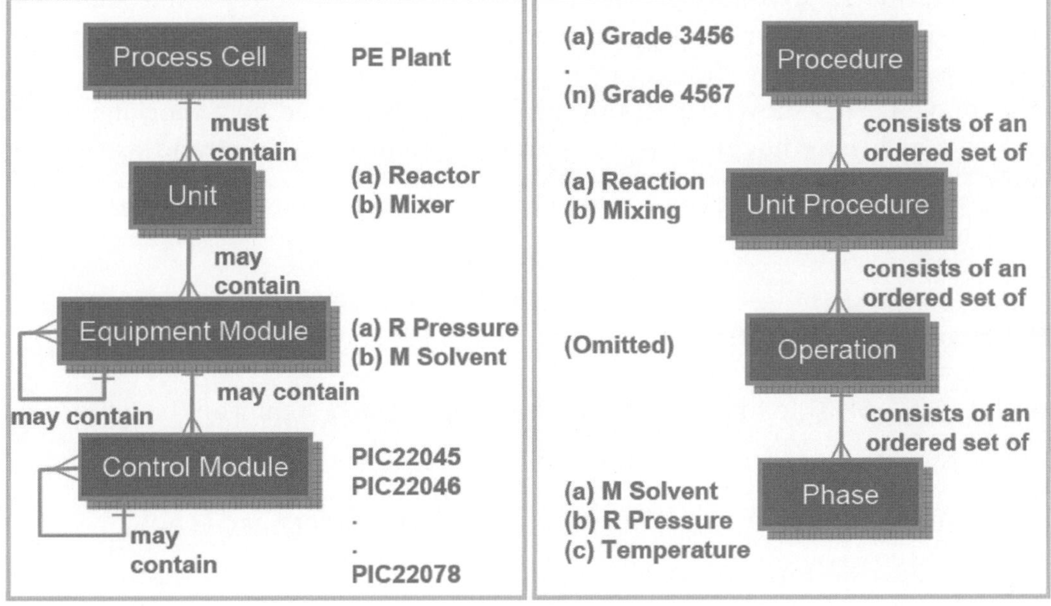

Figure 12.4.   ISA-88.01 structure for batch polymerization.

Traceability and Batch Genealogy

A major item of value to the end user is the ability to trace individual containers of shipped products back to their parent silo batches and then back to identifiable raw materials and conditions in the reactor or extruder at the time the silo batch was resident. This is not as precise as the tracking that can be achieved in the sample batch polymerization process, as time-based calculations have to made on throughput rates with an assumption of linear material movements. In reality, although throughput is directly measured at the extruder, there is a backup of material between the reactor and extruder and product can get trapped at certain points, which is difficult to calculate.

An example of where this is useful is in process investigation. If a quality problem occurs and is detected at the final customer, the container shipment can be traced to the corresponding silo batch, and with a calculated time, traced back to the reactor conditions at the time it was resident. The impact of reactor conditions can therefore be linked to quality, consistency, or other product characteristics. Similarly, if an error is made during the mixing of a batch of material C and detected after testing but before shipment, the affected silo batches can be identified and quarantined or downgraded to a product grade with a lower specification.

In Figure 12.5, container 11521 can be traced to silo 14 and identified batches or lots of materials. By automatically applying a calculated offset from the degassing residence times, a trend of reactor temperature and pressure can be displayed. This shows a pressure peak possibly caused by a blockage in the reactor, which may have impacted quality. Container 11520 straddles two silo batches and so has a wider relevant time slice, covering two batches of material C and two lots of material D.

Summary and Benefits

Applying ISA-88.01 to this continuous process produced significant benefits in the following areas:

- Improved communication via a common vocabulary and set of concepts

- Increased structure and uniformity through the use of ISA-88.01-based batch automation software products in place of custom software developed specifically for the project

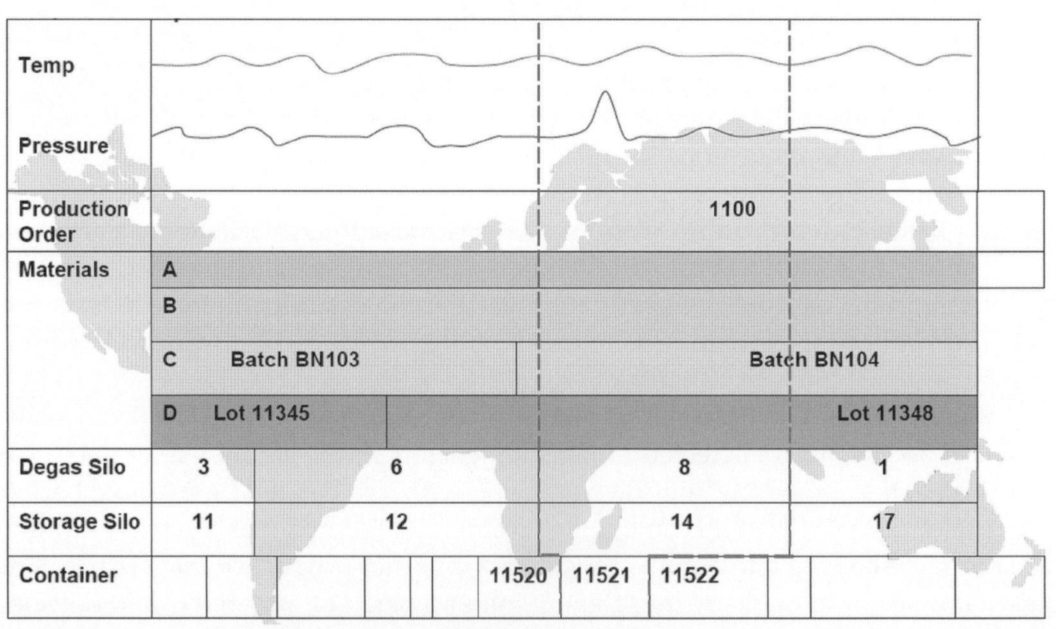

Figure 12.5.   Tracing continuous polymerization.

- Reduced testing time and rapid system development through the use of ready-built ISA-88.01 structures and objects

- Increased user acceptance of the HMI via the use of highly developed visual objects

- Reduced levels of customization compared with that of similar automation systems, allowing the system to be maintained by local support resources

- Increased functionality, with all automation execution taking place at the controller level

- Simplified batch traceability and genealogy, which is particularly valuable to the end user

ISA-88.01 was not applied to the end of the process, where the adoption of even the minimum hierarchy was felt to be too cumbersome. This was particularly true for material transfers between simple storage units where no chemical processing took place. Also, processes in which materials are continuously injected into a batch (rather than added sequentially) are difficult to align with the standard.

Case Study of Upgrading and Adding ISA-88.01 Features to a Legacy System

Presented at the WBF
North American Conference,
Philadelphia, PA,
March 24–26, 2008, by

Robert Durscher
Process Control Systems Engineer
Robert.Durscher@lanxess.com
Rubber Chemicals, LANXESS Corporation,
1588 Bushy Park Road,
Goose Creek, SC 29445, USA

Ken Keiser
Senior Marketing Specialist
Ken.keiser@siemens.com
Siemens Energy and Automation,
1201 Sumneytown Pike,
Spring House, PA 19477, USA

Abstract

This chapter explores the benefits and financial cost justifications of adding an ISA-88.01 batch system to a legacy Distributed Control System (DCS). An ISA-88.01 structured batch management system was added to two different existing legacy control systems. This resulted in one unified Human Machine Interface (HMI) and batch management system managing both legacy systems' controllers and thereby reducing engineering and validation costs.

The plant contained outdated hardware and software in two areas that were controlled by two different control systems. Additionally, both batch management systems did not communicate with each other. The existing HMI software and hardware was also reaching the end of its life for both systems. The plant opted to change the HMI and batch management level for each system to a unified system, rather than replace the entire system. This allowed for keeping the existing controllers and batch code, saving considerable amounts of engineering and validation costs.

Introduction

Most people would say, "It can't be done," when they hear that a legacy DCS is 15 years old and needs to be modernized and fitted with a new HMI from a different vendor than the one that supplied the original system. Throw in the fact that the legacy system is a vintage and code-intensive batch system, and the disbelief intensifies. However that is precisely what happened with a legacy system in a chemical plant in South Carolina. This chapter will describe the steps taken to modernize that legacy system in a stepwise fashion in order to maximize the life of the components, while at the same time improving the overall efficiency by adding ISA-88.01 batch concepts and technology.

History

The plant was opened in 1978 and manufactures high-quality chemicals for the rubber industry. The first DCS was installed in the early 1980s. As was typical of systems during that time, the batch software was rudimentary and not compliant to the then nonexistent ISA-88 standards. The original system looked like Figure 13.1, as supplied by the original vendor (Vendor A).

Problem

Between 1978 and 1999, the system worked well. However, external forces threatened the longevity of the DCS. The batch configuration was not based on ISA-88 standards, which led to inefficiencies when the recipes needed to be changed, and the hardware of this original HMI (herein referred to as HMI "I") was getting more expensive to maintain, causing concern that the system integrity was at risk.

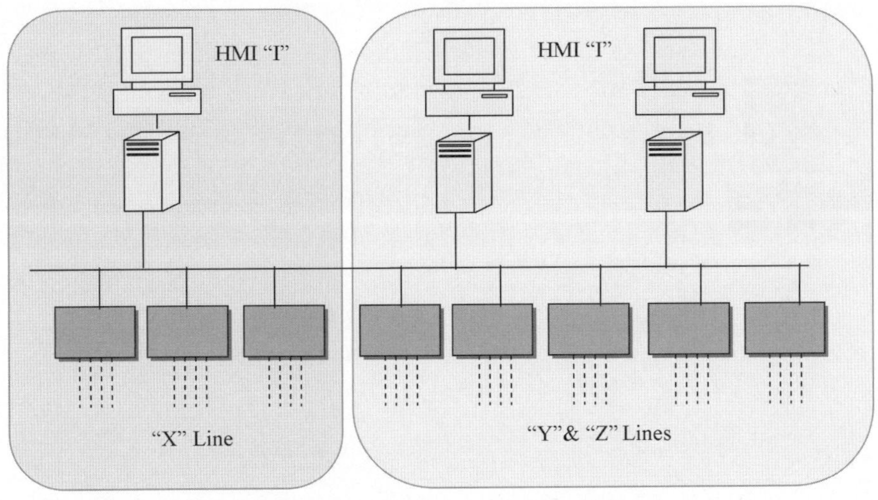

Figure 13.1.   Original architecture of the system from Vendor A.

Modernization

In 1999, a filter was replaced on the "X" line that required either a new controller or a reconfiguration of the existing controller. An analysis was done to determine the best course of action, and based on that analysis it was decided to select a new vendor (Vendor B) that could supply a new controller with an ISA-88.01 solution for that part of the plant. The other parts of the legacy system remained to control their respective areas. After the initial change over to Vendor B's system, the architecture looked like Figure 13.2.

The new control unit and second HMI (HMI "II") proved to be meeting the challenge of running the batch using an ISA-88.01 architecture, and it was decided to continue changing out the other two controllers in the "X" line, along with HMI "I." The final step in the "X" line migration occurred in 2004 and is shown in Figure 13.3. The plant now was automated using two control systems—one from Vendor B and the original from Vendor A. At the same time, the plant recognized that under the current architecture there was a requirement for separate HMIs for the plant process lines. After some investigation, an Object Linking and Embedding for Process Control (OPC) server package was discovered that would allow connectivity to the legacy control system. After installation of the OPC server package, the HMI "II" was able to communicate with the legacy system for continuous points via a gateway as shown in Figure 13.3, but the batch manager was not yet integrated with the legacy system. At this point, operators could monitor

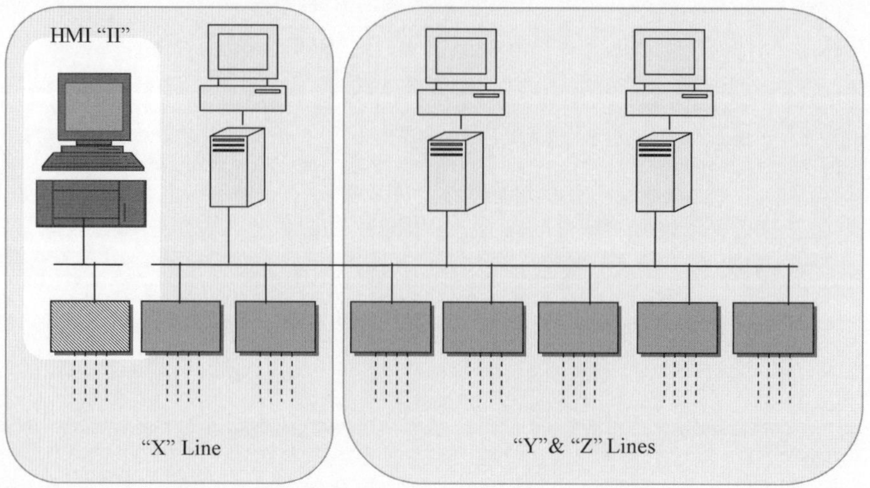

Figure 13.2. Architecture after replacing the first controller with a new ISA-88.01 compliant unit and adding a new HMI "II" with batch management.

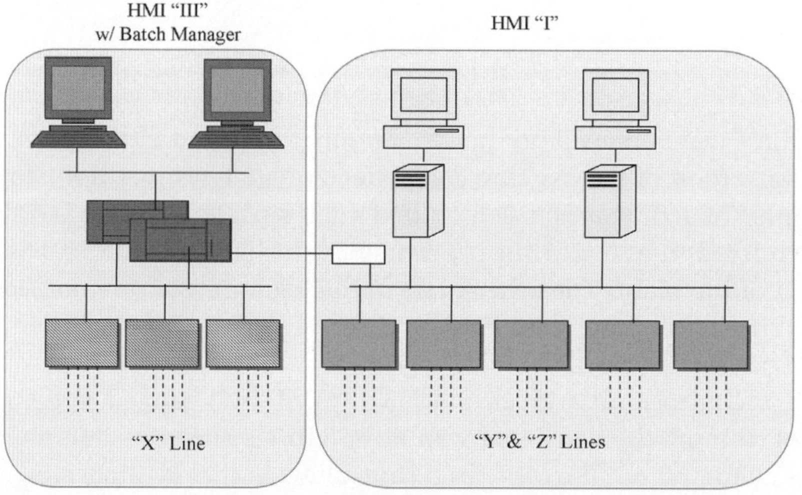

Figure 13.3. Architecture after replacing the entire control system for the "X" line.

all plant processes but required separate HMIs located in the same control room to initiate batch control.

There were issues with the life cycle cost of keeping the older system. The technology was now more than 20 years old, and the following constraints were observed:

- *High maintenance cost*. The replacement keyboard of the HMI "I" cost $3000 USD.

- *Obsolescence risk.* The HMI "I" was due to be phased out of production and support, leaving the plant vulnerable to unplanned downtime.

- *Continued reliance on non-ISA-88 standard procedures.* The technical constraints of the system could not provide for ISA-88 hierarchies and flexibility.

Furthermore, there were advantages to using a modern system in the "X" line:

- Faster configuration time versus the old system, thus saving engineering hours

- ISA-88.01 structure allowing for more flexible batch recipe changes, leading to faster production runs and less time to first batch

- Total cost of ownership was less due to the use of more open, scalable components.

The Idea to Upgrade and Migrate

Continuing the modernization led to the obvious option for the "Y" and "Z" lines, and these were to replace the system in a manner similar to the "X" line. This replacement option required a large engineering investment because the controller code needed to be rewritten. Also, Vendor B had been acquired by Vendor C and announced that the availability of the HMI "II" that was installed in the "X" line would be stopped in 2007 before the controllers and HMIs were replaced. The requirements and feasibility of using the new ISA-88-compliant batch management system for the legacy controllers needed to be determined. The plant engineers determined the requirements for taking the discrete signals and converting them to the integer signals required by the legacy DCS. Testing batch functionality was begun and proven successful in 2004. With the batch functionality proven, it was now possible to work with Vendor C toward a solution that would allow for the use of the third HMI (HMI "III") with all plant controllers. This option would deliver the benefits of ISA-88.01 batch management without having to change the existing controller hardware and configuration on the "Y" and "Z" lines, saving the engineering invested in those controllers over the years. The new HMI would also contain a batch management system that would take the batch and recipe information from the legacy controllers and encapsulate that information into the new batch management system, so that the operators could control the batches from any controller on a unified HMI system. This new HMI "III" would use OPC to communicate to the old system's controllers. Vendor C's solution for HMI "III" also provided several engineering tools to automatically populate the database of the HMI "III" and batch manager, saving migration costs.

The first step (Fig. 13.4) was to upgrade the HMI "II" that was installed in the "X" line. This was a simple upgrade using automatic tools to migrate the database from the original Vendor B's HMI "II" to the upgraded version from Vendor C. The second step (Fig. 13.5) was to install that same version of HMI "III" on the "Y"

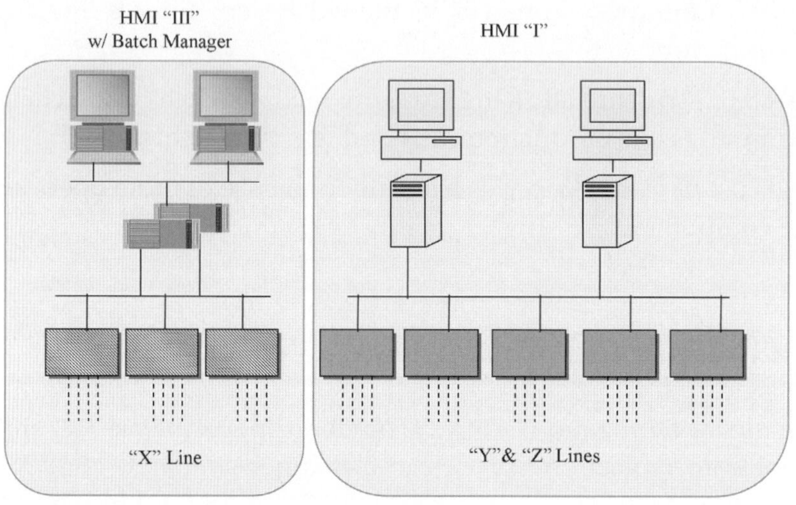

Figure 13.4.   Architecture of the upgraded system.

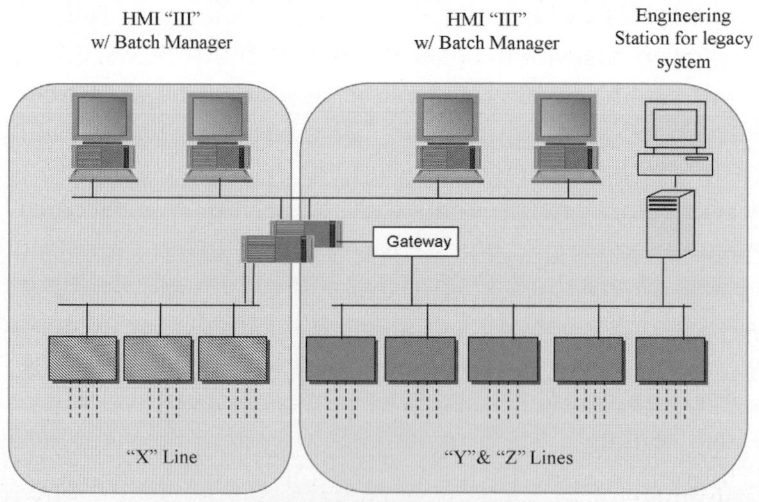

Figure 13.5.   Architecture of the unified HMI with batch management over both the newer and legacy controllers.

and "Z" lines but leave the controllers in place. Figure 13.5 shows the resultant architecture of this unified, ISA-88.01-based batch system. The legacy engineering station was retained to maintain configuration of the legacy controllers as needed.

Methods

The key to the successful migration of this system was based on the software tools included in the newest HMI "III." The following were the steps required to move to the new HMI over the legacy controllers:

- Retrieve the tag database from the legacy system
- Retrieve the batch information from the legacy system
- Import the data (both tags and batch) into the migration tool
- Automatically create symbols and faceplates on the new HMI "III" using the migration tool
- Manually tweak the automatically generated graphics to emulate the old graphics
- View and manipulate the legacy system's points and recipes at the new HMI "III"

Benefits

The following list describes the benefits acquired from the successful integration of the systems:

- The use of ISA-88.01-structured batch concepts and a unified batch management system reduced engineering hours and allowed for graphical monitoring of legacy batches, changes to legacy recipes, and modification of batch parameters while running a batch.
- Continued use of legacy controllers lowered total cost of ownership of the legacy system by continuing to use the existing non-ISA-88 batch and continuous configuration that was developed over 20 years.
- Because of the automatic nature of the migration tools, the new system minimized the amount of manual intervention, thereby reducing engineering hours and maximizing the accuracy of the new HMI "III" and batch manager's database.

- The use of a common HMI reduced real estate needs in the control room and unified the alarm system, which made it more efficient and safer for the operators to control the plant.

- With HMI "III" and a ISA-88.01-compliant batch management system in place, the path was paved for a fully compliant system when upgrading other legacy controllers.

Lessons Learned

In this project, as in all projects, there were steps that should have been taken but were not and the other way around—steps that we took that were unnecessary. However, this project was successful in that most of the time, the right things were done at the right time. The following list describes some of the lessons we learned:

- Use a vendor that is willing to listen to your needs

- Use a standard migration product (rather than a one-time solution)

- Use a local integrator with a knowledge of the legacy system

The next step at our plant is to upgrade the remaining legacy controllers with state-of-the-art controllers that are designed with ISA-88.01 concepts, which are part of the same product line as the current HMI "III." There are now more options for upgrading available from Vendor A, Vendor C, or another vendor entirely.

In any intensive migration effort, it is important to have a partner. In our case, Vendor C listened to our requirements for a unified HMI and batch system, helped us with the startup, and responded to requests for changes. In addition, we had a local integrator who was also the vendor's solution partner, which gave us a company that was located near the plant and knew the legacy system as well as the new system.

Design of a Batch Process Control Tool on the Programmable Logic Controller Platform

Presented at the WBF
North American Conference,
Philadelphia, PA,
March 24–26, 2008, by

Giovanni Godena
Applied Researcher
giovanni.godena@ijs.si
Jozef Stefan Institute, Jamova
39, Ljubljana, 1000 Slovenia

Igor Steiner
Process Control Project Manager
igor.steiner@inea.si
INEA d.o.o., Stegne 11, Ljubljana, 1000 Slovenia

Janez Tancek
Project Engineer
janez.tancek@inea.si
INEA d.o.o., Stegne 11, Ljubljana, 1000 Slovenia

Marko Svetina
Sales and Marketing Manager
marko.svetina@inea.si
INEA d.o.o., Stegne 11, Ljubljana, 1000 Slovenia

Abstract

A new concept has been developed for a tool that allows for the recipe-based control of batch processes according to the ISA-88.01 standard, executed on the Programmable Logic Controller (PLC) platform without Personal Computers (PCs), coupled within the real-time control loop. The concept supports the simultaneous execution of multiple recipes with automatic allocation of units. It also supports parallel and selective branches and enables the execution of the state-transition algorithm for individual phases, extended by the notion of superstates to improve the abstraction level. Most importantly, the concept allows the relocation of the execution of recipes from the inherently unreliable PC platform to the considerably more reliable PLC platform and simplifies recipe-based batch process control systems without necessarily reducing the expressive power and the abstraction level; these are the main benefits of the new concept.

Introduction

Today there is a broad consensus among automation equipment manufacturers, service providers, and end users about the need for compliance with the ISA-88.01 standard[1] for batch process control, regardless of the level of automation of this control. However there are also certain problems associated with the application of the ISA-88.01 standard and any batch tools based on it:

1. Unreliability of the PC platform and real-time communication loop (which can be bypassed through redundancy but incurs significant extra costs)

2. Poor adjustment of the existing tools to the majority of small- and medium-sized projects

3. Unsatisfactory time behavior of the PCs (concerning execution speed and time determinism)

4. Low abstraction level of the phase behavior model, which results in (a) a frequent need to memorize the current step for state transitions that return back to the same step, which is a nontransparent low-level concept and as such a frequent source of errors and (b) complicated logic of individual states, which also often behave as state machines

As a consequence of some of these problems, the contractors of batch process automation projects often decide on different solutions for recipe execution on the

PLC platform, which are mostly oversimplified and for this reason lose most of the expressive power offered by ISA-88.01-based tools. The most frequent example of this is the execution of a recipe as a simple step sequence with a single Procedural Control Entity (PCE) in an individual step, containing all the functionality needed at that step. In this case the only theoretically possible flexibility is in changing the sequence of the PCEs, which in most cases would result in an unfeasible sequence.

This problem can be solved in two ways: The first way is to attain flexible behavior of ample PCEs through their parameterization. The second way is to apply a recipe composed of small PCEs (i.e., phases) with only a single phase per step appearing in the recipe, while other processing is simultaneously performed in PCEs executed separately from the recipe. The problem with this second solution lies in the difficulty of attaining synchronization between the main thread phases and concurrent PCEs. In practice this is mainly solved with hard-coded interdependence, which presents an obvious example of inadequate (too high) coupling and is a great potential source of error. In the rest of this chapter, we describe a batch process control tool on the PLC platform that is designed to avoid the previously mentioned problems as much as possible.

This chapter is structured in the following manner: The section titled "Requirements and Limitations" gives the requirements and limitations that were to be considered in planning the approach. The section named "Concept of a Tabular Recipe" describes the concept of tabular presentation of Sequential Function Chart (SFC) recipes. The next section ("Entity Behavior Model") explains the entity behavior model, and the last section of the chapter presents a short summary.

Requirements and Limitations

Considering the starting points presented in the introduction and the limitations of the target implementation platform, the new batch process tool should satisfy the following requirements:

1. The tool should be executed on PLC and PC platforms. In fact all
   real-time control functions, including recipes, should be executed
   on PLC platforms, and only non-real-time functions such as the
   Equipment Editor, Recipe Management, Batch Management, and
   Archiving should be executed on PC platforms.

2. Recipes should be represented by SFCs or in a sort of tabular notation. The tabular notation would probably be easier to implement
   on the PLC platform than the graphical SFC notation, so we decided

that a table would probably be best. Certain experiences from past batch control projects with tools designed according to the ISA-88.01 standard also made us think in this direction. In one particular project, operators experienced difficulty trying to understand and master the SFC notation but were able to understand the notation when we explained the behavior of SFC recipes to them using tables. Thus we started to think that the tabular notation might in some cases be more acceptable for operators than the SFC notation.

3. Recipe hierarchy should be reduced due to platform limitations. We decided on two recipe levels: a procedure level that contains unit procedures and a unit procedure level that contains phases. By leaving out the operation level, we sacrificed some reuse and a part of the structure, but we estimated that this would still be sufficient for the majority of real (small and medium-sized) projects where the new tool would be used.

4. The behavior of procedural control entities should comply with the ISA-88.01 standard and also include certain model improvements (extensions). Based on our many years of experience modeling procedural control of continuous and batch processes, we reached the conclusion that the proposed phase behavior model was too simple, and consequently, the content of individual elements of the model were too complex.

5. A simple but efficient state propagation mechanism among procedural control entities should be determined for this tool. The state and mode propagation mechanism of batch control tools is one of their more complex aspects. Thus, extra attention should be paid to the propagation mechanism during development.

6. The tool should be designed in manner that achieves the highest possible level of portability (to various PLC platforms).

Concept of a Tabular Recipe

The concept of a tabular recipe is based on mapping the SFC recipe design to a table and executing this table on the PLC via the table interpreter. The tabular record of recipes is referred to as Tabular Batch Recipes (TBR). The shortest possible description of the basic operational concept of the TBR interpreter is as follows: *the condition for recipe transition to a new row requires the termination of all dominant phases of the current row that are not repeated in the next row as the same instance.* This definition requires further explanation before describing the SFC to table mapping rules and the table execution algorithm.

System Phase Parameters

Two system phase parameters are needed for the execution of tabular recipes or step transition of these recipes; these parameters define phase dominance and phase instance. The first parameter defines phase dominance; therefore it defines whether the termination of a phase represents a condition for recipe transition to a new row. As a rule it also indicates if the phase has its own (intended) completion (e.g., dosing a material) or if it is inherently designed to be executed infinitely (e.g., mixing). If the phase is dominant, the recipe row transition is possible only after termination of the phase. If the phase is dependent, the recipe row transition happens independently of this phase, and if the phase is active at that time, the recipe interpreter stops it.

The dominance attribute value of an instance of an inherently dominant phase can be changed to dependent behavior (i.e., we can take off the phase the decision on its completion and pass it to the recipe interpreter), but the opposite does not hold true; an inherently dependent phase can under no circumstances obtain dominant behavior, since it is not programmed to behave in that way.

The procedure for setting the value of the dominance attribute for phase instances is as follows: The recipe editor during the creation of a recipe copies the phase's own (default) value of the dominance attribute from the Phase Control Block (PCB) to the dominance attribute for an instance in the recipe. If the default value of phase dominance (in a PCB) is "FALSE," the value of dominance for phase instances cannot be changed in the recipe. Therefore the default value is also the only possible value. However, if the default value of dominance in the PCB is "TRUE," the value of the dominance attribute for phase instances in the recipe can be changed during editing.

The second system phase parameter, named Instance, defines the instance of an individual phase in a recipe. This parameter is needed when the same phase appears in several consecutive recipe table rows. In this case, the phase behavior on row transition depends on whether the instance of this phase in a new row is the same as its instance in the current row (which indicates the same meaning as the appearance of this phase in a long thread in the SFC) or if there is another instance of this phase (which indicates the same meaning as the repetition of the same phase in several consecutive steps of the SFC).

Mapping an SFC Recipe to a Table

Figure 14.1 shows an example of a recipe, and Table 14.1 shows its mapping to a table. In this example the dominant phases are marked with an asterisk (*). The algorithm for mapping the SFC into the table is the following: We move along the SFC thread that contains the most phases. Each of these phases is mapped into a distinct table row, and all phases that appear concurrently in parallel SFC threads are mapped into the same row. Therefore, the mapping of our example is the following:

- Phase A has no concurrent phase, and therefore it is mapped alone into the first table row.

- Phase B has a concurrent phase, G; therefore these two phases are mapped into two cells of the second table row.

- Phase C has two concurrent phases, D and G; therefore these three phases appear in three cells of the third table row.

- Phase E also has two concurrent phases, D and G; therefore these three phases appear in three cells of table row four.

- Phase F has a concurrent phase, G; therefore these two phases are mapped into two cells of table row five.

- Phase H has no concurrent phase, and therefore it is mapped alone into table row six.

- Phases I and J are concurrent and are mapped into two cells of table row seven.

- Phase I appears alone (without any concurrent phases), and therefore it is mapped alone into table row eight.

Behavior of the Table Interpreter

The execution of the recipe table is performed rather intuitively. For the considered example (Table 14.1), it runs as follows:

- In the first row there is only one dominant phase, A; therefore the termination of this phase implies the recipe row transition.

- In the second row there are two dominant phases, B and G, but only the termination of phase B implies the row transition because phase G is repeated in the next row as the same instance. During row transition the interpreter does not influence the execution or the state of phase G.

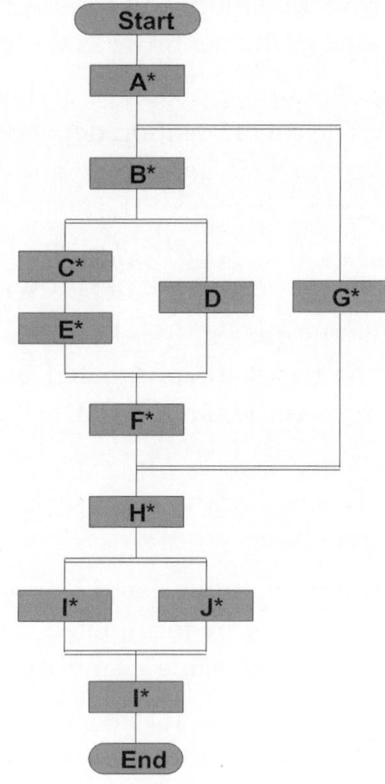

Figure 14.1.   SFC recipe.

Table 14.1. Recipe table		
A*		
B*	G*:1	
C*	D:1	G*:1
E*	D:1	G*:1
F*	G*:1	
H*		
I*:1	J*	
I*:2		

- In the third row two dominant phases, C and G, appear as well as one dependent phase, D. The row transition depends only on the termination of phase C, as phase D is a dependent phase and phase G is repeated in the next row as the same instance. The state of phase

G is not affected by row transition, and the same applies to phase D, as this phase is repeated in the next row as the same instance.

- In the fourth row two dominant phases, E and G, appear as well as a dependent phase, D. The row transition depends only on the termination of phase E, as phase D is dependent and phase G is repeated in the next row as the same instance. The state of phase G is not affected by recipe row transition, whereas phase D is stopped by the interpreter (as it is not repeated in the next row).

- In row five two dominant phases, F and G, appear. The row transition occurs when both phases are terminated, as neither of them is repeated in the next row as the same instance (in this case they are not repeated at all).

- In row six there is only one dominant phase, H; therefore its termination causes the row transition.

- In row seven two dominant phases, I and J, appear. The row transition occurs when both phases are terminated, as neither of them is repeated in the next row as the same instance.

- In row eight there is only a dominant phase, I; therefore its termination causes the row transition.

Possible Structures of Recipes

The structure of a recipe is a two-level structure—the procedure owns its recipe (i.e., its higher-level table), which contains unit procedures, and each unit procedure owns its recipe (i.e., its lower-level table), which contains phases of this unit. The only limitation of recipe structure at both levels is that conditions must be provided for the transition to a new step, which means that at any point of the recipe only one of the currently active concurrent threads contains more than one step (and a step can be a phase or a new concurrent branch). This limitation has no consequences in most real cases, but in the remaining cases described in this chapter, consequences are shown in the form of a partly reduced concurrency in the recipe; in other words, the execution of individual recipe parts can be slowed down.

Handling Recipe Formulas

The current version of the tool supports recipe phase formula value parameters, which are specified in the Recipe Editor for a recipe when the recipe is created. Other ways of configuring phase parameters are currently under development and will be available in the near future. One of these methods allows operators to specify a parameter value during the run time, prior to executing the recipe phase or using the existing recipe parameter.

Specific Transition Conditions: "AND" and "OR" Recipe Branching

The presented tabular notation of the recipe that it supports "AND" branching and transition conditions based on the termination of given (dominant) phases. In the recipe system, we certainly also expect the option of "OR" recipe branching and specific conditions for recipe transition, such as reaching a defined value of a given process variable. These and similar requirements can be satisfied with the introduction of special phases. Here, we will introduce two such special phases: phase Condition and phase Branch, since what we are doing is called a "Conditional Branch."

Special Phase Condition

The special phase Condition serves for the implementation of situations when a logical expression performing a relational operation on a given process variable appears as a condition for step transition in the SFC recipe. In TBR this is solved by the special phase Condition, which repeatedly performs the relational operation, and when the comparison result becomes "TRUE," the phase completes. In our model, if this phase was the only dominant phase in the row where it was placed, we attained the desired recipe behavior. The step transition will occur when the condition is satisfied (i.e., at "TRUE" value of the relational operation).

In the Condition phase the comparison type (relational operator) and comparison objects (what is compared with what) need to be stated. We certainly want this phase to be generalized; we do not want to introduce a distinct Condition phase for each process variable that we want to introduce into a recipe for comparison. This can be achieved with the corresponding phase parameterization and the following parameters:

- First operand type [variable | constant]
- Id/value of the first operand (considering whether it is a variable or a constant)
- Relational operator [= | <> | < | > | <= | >=]
- Second operand type [variable | constant]
- Id/value of the second operand (considering whether it is a variable or a constant)

Special Phase Branch

The special phase Branch serves for the implementation of "OR" recipe branching, when the execution of the recipe is threaded in two possible directions considering the given condition (the value of the logical expression, as in the Condition phase). Such behavior of the Branch phase is attained by enabling the phase write access to the system variable containing the value of the recipe step. The phase decides based on comparison, and therefore it has the same five parameters as the Condition phase and two additional parameters:

- Next step for comparison result "TRUE"
- Next step for comparison result "FALSE"

Using the Branch phase, the correct structure is not provided by the recipe editor as with the SFC graphic editor, but the recipe creator must pay attention to it (as is the case with programming languages lacking structured programming constructs). We can also implement the Jump special phase, which performs an unconditional change of step in the execution of the recipe. This phase has only one parameter—Next Step.

Entity Behavior Model

PCE states arise from the behavior model as it is proposed by the ISA-88.01 standard. The model is extended in certain elements to attain better properties, particularly those that increase the simplicity and reliability of the application software.

Phase Behavior

In order to attain the simplest phase logic building blocks possible, it is essential to build a model where phase building blocks are merely simple sequences, without any internal states or threading, that do not require individuals to memorize the entry point to the sequence (e.g., whether we came to the Running ENTRY sequence from state Idle or from state Restarting, as actions in these two cases differ). In compliance with this and the behavior model of procedural control entities in our domain-specific modeling language ProcGraph,[2] we defined the phase behavior model as an extended state machine model, as shown in Figure 14.2. The extension is comprised of the following properties:

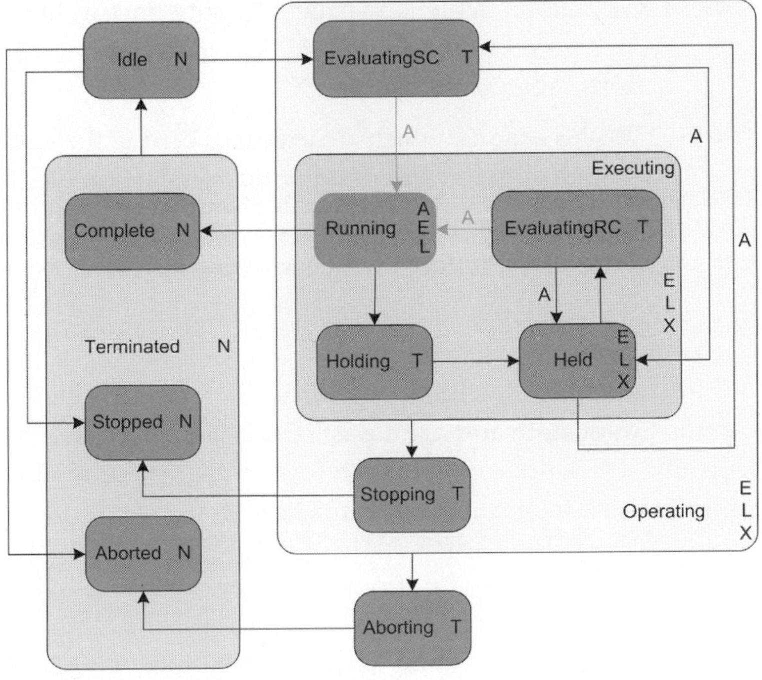

Figure 14.2.   Phase state transition diagram.

1. Introduction of new elementary states: Evaluating Starting Conditions (EvaluatingSC) and Evaluating Restarting Conditions (EvaluatingRC) to achieve distinction among holding causes, phase initial conditions, and phase restarting conditions

2. Introduction of nested states (elementary states and a number of super states)

3. Introduction of a very fine granularity of the processing (action) structure

    a. State sequences of actions. There are five different possible sequences for each state

      ■ ENTRY actions, which are executed only once on entry to a given state (in Fig. 14.2 the states having ENTRY actions are marked with an "E")

      ■ LOOP actions, which are executed cyclically for the entire time the phase is in a given state (in Fig. 14.2 the states having LOOP actions are marked with an "L")

- EXIT actions, which are executed only once at the exit from a given state (in Fig. 14.2 the states having EXIT actions are marked with a "X")

- ALWAYS actions, which are executed during the state activity and during transitions entering this state (in Fig. 14.2 the states having ALWAYS actions are marked with an "A")

- TRANSIENT actions, which are executed only once at transient states (in Fig. 14.2 the states having TRANSIENT actions are marked with a "T")

b. Transition actions, which are basically specific ENTRY actions of the target state and are a part of this state executed before the ENTRY actions of this state (in Fig. 14.2 the transitions having actions are marked with an "A")

Actions of all sequences have durations (i.e., they are not instantaneous). This applies not only to ALWAYS actions or LOOP actions, as is the case in the Statechart notation,[3] but also to ENTRY, EXIT, and TRANSIENT state actions and transition actions. The advantage of this is that there are no problems involved with synchronizing individual activities. ENTRY and LOOP actions are not executed at the entry to a state if the condition for termination is satisfied or if any failure cause is present, but EXIT actions are executed in any case.

State and Transition Types

Our phase behavior model is comprised of different state types that are classified according to two criteria. According to the processing criteria, states are divided in the following way:

- Quiescent states are states without any processing, marked with an "N" (left part of Fig. 14.2).

- Active states are states that contain certain processing (right part of Fig. 14.2).

According to the duration criteria, states are divided in the following way:

- Transient states are those states that contain only one sequence, and when they are executed, a transition to another state occurs. Each active transient state contains one TRANSIENT sequence.

- Durative states are those states in which a phase normally remains for a longer time. The processing of active durative states is divided into several sequences, which can be of ENTRY, LOOP, EXIT, and ALWAYS type.

- Quiescent states (which are all also durative states) are Idle, Complete, Stopped, Aborted, and Terminated.

- Active durative states are Running, Held, Executing, and Operating.

- Active transient states are EvaluatingSC, Holding, EvaluatingRC, Stopping, and Aborting.

Transitions are divided into active and inactive transitions. Active transitions contain certain processing. These transitions are from EvaluatingSC to Running, from Starting to Held, from EvaluatingRC to Running, from EvaluatingRC to Held, and from Held to EvaluatingSC. Each active transition contains one sequence of transition actions.

Complex Transitions

Complex transitions are a series of state and transition sequences between two durative states or between a durative state and a branching transient state, which is a switch that paves the way to one of the two possible target durative states. In other words, complex transitions are integrated processing wholes (composed of a number of sequences) during the execution of a phase state machine. The execution of a phase state machine is performed by the Phase Logic Interface (PLI), which checks for instances where it may be necessary to activate one of the complex transitions that are available for each active durative state. When the conditions (or causes) for the occurrence of a complex transition are met, the PLI executes a complex transition by consecutively activating all sequences of this complex transition.

Phase Logic Sequences

Phase logic sequences present a framework for the modularization of batch control software. Considering the phase behavior model shown in Figure 14.2, the theoretical number of sequences for each phase is fifty eight with four sequences for each durative state; there are nine durative states (elementary state or super state), one sequence for each transient state, five transient states, one sequence for each transition, and seventeen transitions. It is clear that many sequences are not needed, and thus their number can be reduced significantly.

The first "trivial" level of sequence reductions refers to all sequences of quiescent states and transitions between two quiescent states. All these sequences were not needed in the model, and they were excluded. The second level of sequence reduction is executed in the following way: all the sequences that appear together in a complex transition and only in this transition can be replaced with a single sequence. After performing these reductions, the phase contains twenty-two sequences: seventeen in states and five in transitions.

Further Reduction of the Phase Behavior Model

The described phase behavior model with twenty-two sequences presents the maximum possible range of elements or sequences and therefore the minimum possible granularity of the execution of phases. In case a certain programmer (or organization) thinks that the these twenty-two sequences are too many, as he has built a simpler phase behavior model (with a lower number of sequences), he should only configure the sequences that he does not want to use as inactive. The PLI ignores (does not activate) sequences configured in this manner. Theoretically the programmer can also program applications by using only one sequence; although in most cases this would look more like "hacking" than programming. The programmer can also use exactly the same five active sequences that the currently available batch process control tools use in accordance with the ISA-88.01 standard.

Recipe Behavior Model

Just like a phase, a recipe is also executed as a state machine. The essential difference is that a recipe state is composed of two substates—ExecutingStep and AdvancingStep. In the state of ExecutingStep, the interpreter first resets (puts in state Idle) all phases of the current step. Then it activates them and waits for all conditions to be met for the transition to a new row in the recipe table. When this happens, the interpreter passes to the AdvancingStep state, where the current step is "cleared" (i.e., all dependent phases that are not repeated as the same instance in the next step are stopped by the interpreter's command).

During particular recipe states, the interpreter influences the phases of the current step by sending them commands. For example, during the recipe state ExecutingStep the interpreter sends commands Reset and Start to phases, and during the recipe state AdvancingStep it sends the command Stop. During recipe state Holding it sends the command Hold, during recipe state Stopping it sends the command Stop, and during recipe state Aborting it sends the command Abort. Figure 14.3 shows the recipe state transition diagram.

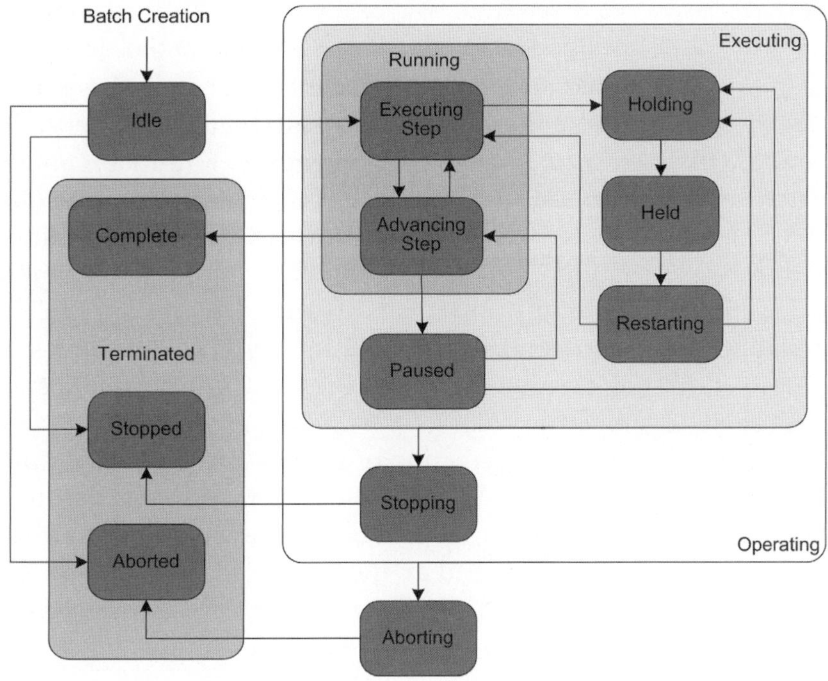

Figure 14.3. Recipe state transition diagram.

State Propagation Mechanism among Procedural Control Entities

Batch process control tools in accordance with the ISA-88.01 standard that are executed on PCs have integrated state and mode propagation mechanisms for procedural control entities. A similar but simpler holding propagation mechanism is also integrated in the new tool concept. In higher-level recipes that are composed of unit procedures, recipe holding (due to a command) is propagated downward to all unit procedures; however, holding of individual unit procedures is not propagated upward to the recipe. In lower-level recipes there is a propagation mechanism that is executed through two (system) phase parameters: one parameter defines whether the holding of the phase is propagated upward to the recipe containing the phase, and another parameter defines whether the holding of the recipe containing the phase is propagated downward to this phase. The proper combinations of these two parameter values in individual phases can enact various behaviors: when a defined phase is held, the recipe may or may not go to holding, and when the recipe is held, certain phases go to holding while others remain running.

Summary

A new concept was developed for recipe-based control of batch processes according to the ISA-88.01 standard. Usually recipes are created on a batch server (normally an industrial PC) that acts as a batch execution engine. Within a real-time control loop the batch server communicates with a PLC during recipe execution. Phase logic and basic control of the process are performed by the PLC. Since PCs are inherently unreliable, redundancy of the execution engine and communication lines is used to increase the safety and reliability of the operation. PLCs are one of the most reliable equipment components within a production process. They are becoming more and more capable. Based on these facts, a new concept for batch execution has been developed.

Within the new approach, the creation of recipes and batches is done on a PC and then downloaded to a PLC. The PLC itself becomes the batch execution engine while still taking care of the phase logic and basic control of the process. The need for real-time communication loops and an industrial PC are eliminated. The need for redundancy is dramatically reduced. The new concept is simple, cost effective, reliable, and understandable, yet sufficiently effective for most batch process control applications.

The concept is based on the mapping of the SFC recipes to a tabular notation. It features simultaneous execution of more recipes with automatic allocation of units, support of parallel as well as selective branches, and the execution of the state-transition algorithm of individual phases.

The phase behavior model is extended by the notion of superstates to improve the abstraction level. As a consequence of the extended behavior model, the phase logic is composed of essentially simple sequences without branches, based on sequence execution history or elements of the state-transition algorithm built into application logic. The model is also capable of remembering the step counter in the running state before holding, so that it can continue the execution at the appropriate point after restarting. An additional benefit of the introduced state-transition model is that the programmer is allowed to select a particular subset of the predefined states and super states in order to define his or her own model of phase behavior.

Based on the presented concept, a batch control tool called PLCbatch has been developed. A beta release is currently available. PLCbatch has been used in the semi-industrial batch neutralization plant in the process laboratory at the Department of Systems and Control at the Jozef Stefan Institute in Ljubljana, Slovenia. In the near future the tool will also be used in the automation of real industrial processes.

References

1. Instrumentation, Systems, and Automation Society. 1995. *ANSI/ISA-ISA-88.01-1995: Batch Control, Part 1: Models and Terminology*. Research Triangle Park, NC: ISA.

2. Godena, G. 2004. Procgraph: A procedure-oriented graphical notation for process-control software specification. *Control Engineering Practice* 12: 99–111.

3. Harel, D. 1987. Statecharts: A visual formalism for complex systems. *Science of Computer Programming* 8: 231–74.

Leveraging the ISA-88.01 Standard in a Multi-station Dipping Process

Presented at the WBF
North American Conference,
Philadelphia, PA,
March 24–26, 2008, by

John Robert Parraga
Process Consultant
jrparraga@ra.rockwell.com
Rockwell Automation,
15458B North 28th Avenue,
Phoenix, AZ 85053, USA

Marcus A. Tennant
Product Manager
matennant@ra.rockwell.com
Rockwell Automation,
1201 South Second Street,
Milwaukee, WI 53204, USA

Abstract

Anodizing and cleaning are critical applications in many industries. This chapter outlines the application of ISA standards to the requirements of a cleaning and anodizing process and describes the following:

- An overview of an anodizing process, typical equipment layout, and the critical control points in the application to ensure quality, consistency, and regulatory compliance

- How the basic ISA-88.01 tenants of separating procedures from equipment would apply in the application

- An outline of the unique equipment arbitration requirements in a multi-tank dipping process

- Movement requirements in the application and the criticality of synchronization between units in the process to ensure accurate and consistent processing times for each application step

- How integration of automated and manual procedural steps within the standards model can improve compliance and quality

- How using the scheduling capability of the automation layer can yield improvements in the efficiency and quality of the system

Industry and Requirements Overview

Throughout many industries, there are often manufacturing steps that require processing parts at a sequence of stations. Some examples of this type of process are as follows:

- *Metal anodizing.* Processes materials through a series of immersion tanks with steps such as pretreatment, anodizing, rinsing with deionized water, and sealing

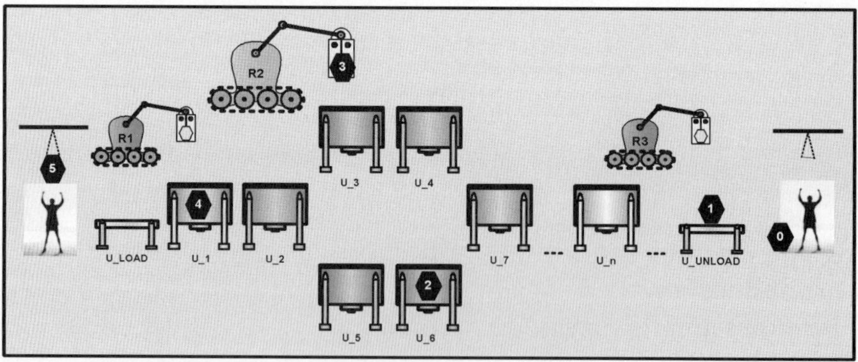

Figure 15.1. Process overview.

- *Material deposition.* Moves electronic or printed components from one station to another for layering of materials to build the circuit or image

- *Parts cleaning processes.* Cleans parts by moving them through a series of baths that may contain caustic chemicals, acid, surfactant, or an ultrasonic or mechanical cleaning device and a rinsing solution

- *Semiconductor wafer tools.* Moves semiconductor wafers through many processing steps that perform a series of tasks to complete a portion of the manufacturing process

- *Chemical milling.* Coats materials with chemical resistant coating and exposes remaining surfaces to etching chemicals to take away (mill) material to a final engineered requirement

Challenges of the Process

In many of these applications, multiple parts are being processed simultaneously and are moved from station to station by a robot arm, robot, or some other transport device or by an individual. In most cases the soaking or processing time at various stations is critical, while the time at other stations may not be as critical. In the manufacturing process, a shortened or lengthened processing time can diminish quality and possibly increase scrap or reprocess rates. Not only must the processing time in critical stations adhere to strict tolerances, but typically and equally important, these steps must be followed by stations that remove or neutralize the remaining critical solutions from the parts. There must be a location for this next step of the process for each component produced, whether they are small semiconductors or very large piece parts for building construction, aerospace applications, or vehicle components.

To solve these scheduling and arbitration problems in a multi-unit dipping and transfer application, custom code is typically written to follow a specific recipe (procedural model, without an equipment model) for each application built. This methodology has inherent life-cycle issues, from control change issues and update issues, to challenges of system expansion and understandability issues that may arise when the primary author leaves the company or retires.

How ISA-88.01 Can Be Applied

The following section describes how an ISA-88.01 batching system controls and schedules the processing of parts in a multi-tank processing (dipping) line. A brief introduction to an ISA-88.01 concept will allow us to understand some of the basic concepts followed in the design and implementation of the batching solution for these types of processes.

One of the most basic and fundamental concepts that ISA-88.01 advocates is the separation of the capabilities of the plant or the physical model from the procedural model—in other words, the separation of "what is the equipment capable of" from "what do we want to do with the equipment." If the plant capabilities are programmed in a modular way, then a recipe or ISA-88.01 procedure can be created that specifies what the equipment should do.

Based on this solution approach, we first focus on "what the equipment is capable of"; this capability is captured in what we like to call the *area model*. At first glance, certain types of processes may not seem to be batch processes. But looking at them from the process standpoint, we discover that if a process consists of a sequence of steps (a procedure) that may be performed in one or many pieces of equipment, it can be considered a batch process. Thus, the process may be broken down in a modular way and improved by applying off-the-shelf products that conform to or follow the ISA-88.01 models.

The following is a list of steps that are typically used to determine the equipment model for a traditional batching process:

1. Determine the boundaries of the project

2. Determine the units and containers

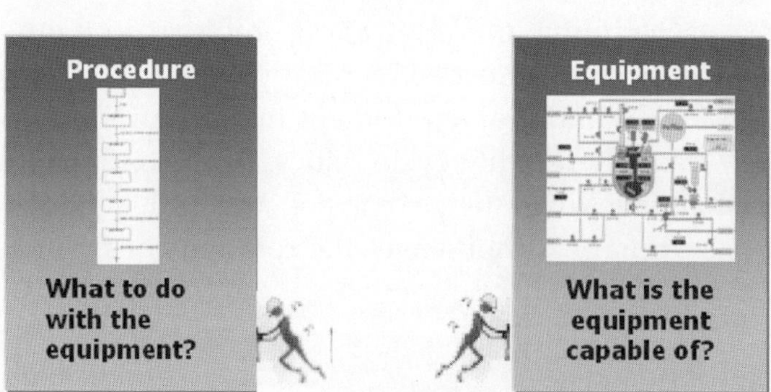

Figure 15.2.   Separation: A fundamental concept.

3. Follow the material flow

   3.1. Bring the material

   3.2. Move the material

   3.3. Remove the material

4. Ask, what is done to the material while in the unit?

5. Identify "unit tags"

6. Determine the equipment arbitration

7. Define the parameters and report values required

After looking at our sample process, the following area model is defined:

1. Each processing tank or station is considered to be a unit.

2. The loading and unloading stations at the beginning and end of the process are considered units.

3. Each unit will expose as unit tags pertinent processing information such as

   - Processing elapsed time,

   - Processing remaining time,

   - Parts present at unit, and so on.

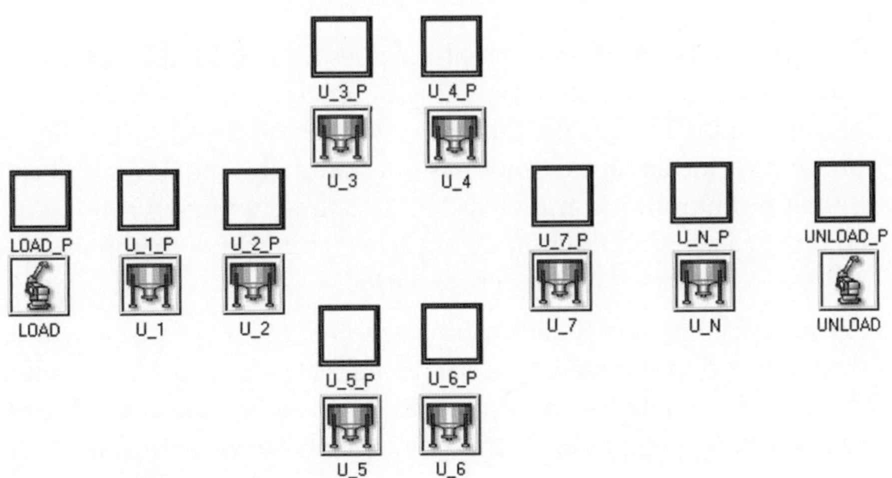

Figure 15.3.   Area model units.

4. Each unit has what we refer to as a phantom unit (which can not be found in ISA-88.01). In addition to the traditional unit, we have created what are considered phantom units of these previous units. The reason for these phantom units is to provide functionality that is required to improve the throughput of the process and work around the material transfer limitations of our system.

5. Each unit typically contains these basic equipment phases

- Transfer IN (X_IN)
- Transfer OUT (X_OUT)
- Processing (these are typically timers that are triggered by various conditions)
- Ready to Receive/Send (Pick/Place)
- Inter-unit Synchronization
- Operator prompts
- Equipment initialization
- Timer

Timer Phase Description

The Timer phase consists of a timer that can be used by the recipe. The phase has a mode, which refers to the action it must take in the event that the phase goes into the Held state and then is restarted. The Timer phase modes are as follows:

1. *Continue.* In the event that the phase goes into Hold, the timer will continue to measure the time. If the timer completes while the phase is in the Held state, the phase will remain Held. After the phase is restarted, if the timer has not timed out, the timer will continue timing until it completes. At this point the phase will complete as well. If the timer completes while Held, the phase will complete as soon as it gets restarted.

2. *Retentive.* In the event that the phase goes into Hold, the phase will stop the timer and the elapsed time is retained. After the phase is restarted, the timer will continue timing from the previously accumulated time until the timer completes. At this point the phase will complete.

3. *Reset.* In the event that the phase goes into Hold, the timer stops the time. The elapsed time is not retained; it is reset to zero. After the

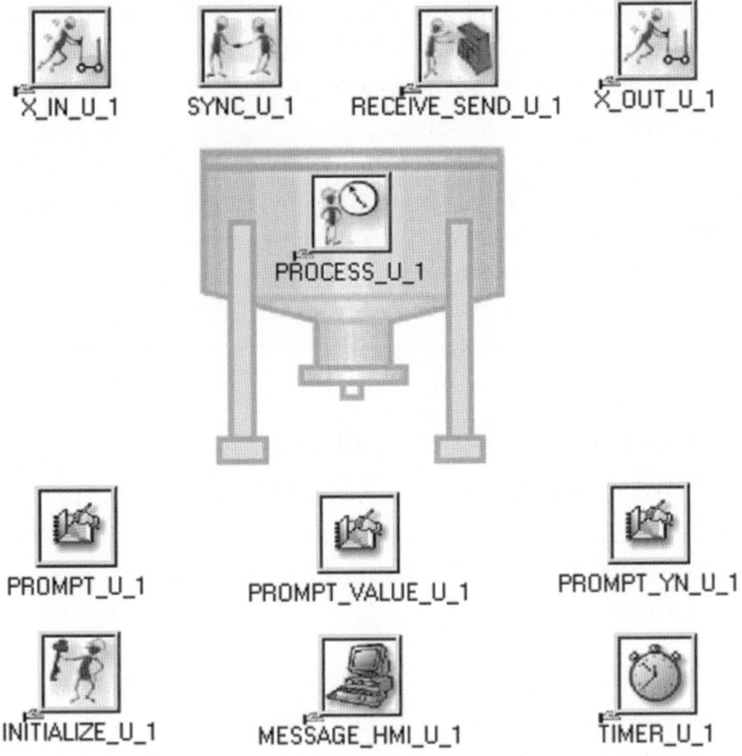

Figure 15.4.   Area model unit 1 equipment phases.

phase is restarted, the timer continues timing until it completes. At this point the phase will be completed.

Initialize Phase Description

The Initialize phase checks the equipment status for all equipment in the formulation tank as well as the skid. It will prompt the operator if any of this equipment is not ready (e.g., in fault or not in auto mode). After acknowledging this the prompt, the phase will recheck the equipment. The phase will not complete unless all equipment is ready.

Process Phase Description

The Process phase is similar to the Timer phase and consists of a timer that, in most units, starts after the parts are placed in the tank. In other cases, the processing timer initiates and tracks processing time based on other conditions such as electrification or deposition time. On the loading station the PROCESS phase is responsible for keeping track of the part identification tags. The unloading PROCESS phase as well as the work stations keep track of the task duration.

Message_HMI Phase Description

The Message_HMI phase displays multiple string messages originated in the recipe. This text is used for the purpose of informing the operator of the progress of the recipe being performed in a unit. The text displayed on the Human Machine Interface (HMI) is a string value that is entered as a parameter in the recipe.

Receive/Send (Pick/Place) Phase Description

This phase indicates to the Robot Equipment Module (EM) if the part being processed is ready for pick (in the source) or if it is ready for place (in the destination). This information is then used by the Robot EM.

Sync Phase Description

This phase is used to synchronize the steps of different units. This phase is intended to operate in conjunction with another Sync phase and must be linked in a link group in the recipe procedure. The example shown in Figure 15.5 demonstrates the use of the Sync phases. The recipe in unit A is dependent on the recipe in unit B. We can only start adding catalyst to unit A if the unit B temperature is greater than XX degrees.

Robot EM Description

In a traditional batch process, the batch or product is pumped or gravity fed from one unit to another; in parts processing types of processes, the batch or parts are transported from unit to unit by means of cranes, robots, or personnel. Similarities can be drawn between the pump and the cranes or robots because they perform the similar task of moving something from one location to another. Therefore parts that transport equipment (fixed or mobile robots) are considered shared EMs whose sole purpose is to receive a command to move parts from a source to a destination. The "Transfer IN" equipment phases will acquire the optimum transport EM during processing.

The task of moving parts from unit to unit (loading station, tanks, and unloading station) is performed by the Robot EM. Figure 15.6 represents a Robot EM. The operator or the program is able to take control of the Robot EM and set up the

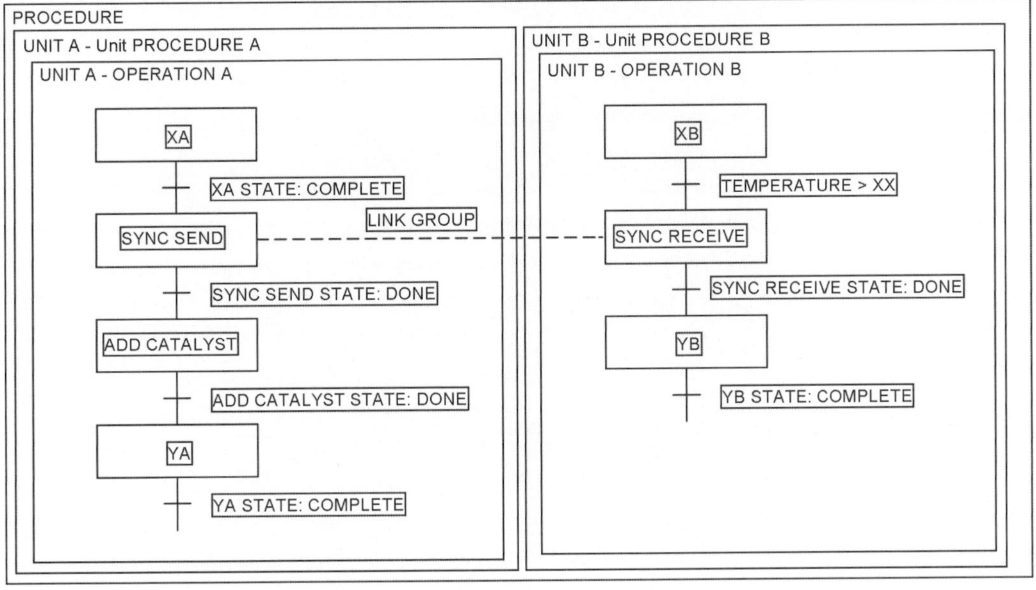

Figure 15.5.   Sync phase example.

appropriate operational parameters, the source, and the destination where a part will be moved.

The robot selection method can be established based on a recipe parameter. The recipe can be set up to use a pre-established robot or it can use different criteria to determine the robot that will be used. Once the phase establishes a robot, it will request acquire that resource via batch server. Once this resource is allocated to the X_IN phase, the phase downloads the appropriate parameters to the Robot EM and commands it to proceed with the transfer.

The robot will move to the requested source and lift (pick) the part only if cleared by the "Source Ready" signal. This signal will be set by the Send/Receive phase if the system is in auto or may be set by the operator if in manual mode. If the Source Ready signal has not been received when the robot arrives at the source, then the robot has the ability to set itself up to retrieve the part without actually lifting it from the source; how far the robot should go is specified as a parameter.

After lifting the part, the robot will deliver the part to the desired destination. If the destination is not ready then the robot can be instructed to wait at a preestablished "destination wait point." Once the "Destination Ready" signal is received, the robot will proceed to the final destination and lower the part and release it, after which it will proceed to raise the hoist and move to a "post destination end point."

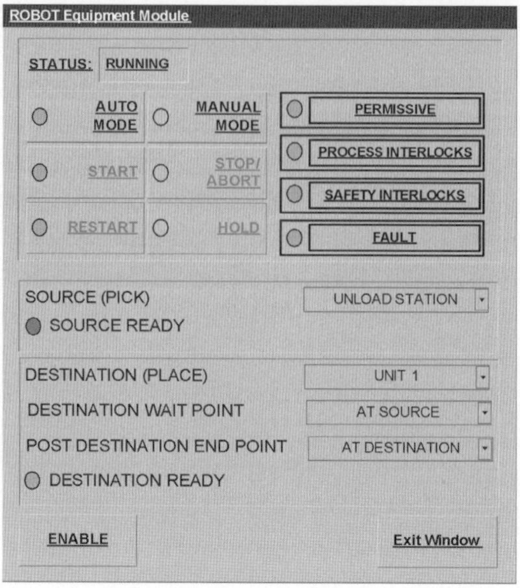

Figure 15.6.   Robot EM.

Control Types

Three types of control are needed in batch manufacturing: Basic Control, Procedural Control, and Coordination Control. The sequencing requirements for a dipping and soaking application are reasonably straightforward. Typically for each operation, there is a "Transfer IN" operation followed by either a "dip," "soak," or "electrolysis" process operation. Following this, the part is removed and transported to the next processing station or to the completed process stage. At first glance, the sequencing requirements do not invoke ISA-88.01 design principles, due to the modularization of the process; it is also not evident that the advantages of separating the equipment from the procedures could be realized.

Among the complex requirements for this application is the Coordination Control between units as well as other pieces of processing equipment that need to be allocated and arbitrated. These types of applications have an overall cross section of control problems.

Basic Control

Dedicated to establishing and maintaining a specific state of equipment and process, Basic Control includes the following:

- Regulatory control

- Monitoring

- Interlocking

- Exception handling

- Repetitive discrete or sequential control

There is a significant amount of basic control in the processing, along with interlocks and permissives. Basic control is used to maintain the tank solution's physical conditions like temperature and level and also used to open and close the tank entry port. Basic control is also used to maintain chemical conditions such as pH and solution concentrations.

Another form of basic control takes place in the robot logic. Each robot is capable of performing the transfer of parts between units; the EM of each robot is also capable of requesting other robots to move to a different location. This is done, if required, to clear the physical path for the requesting robot.

Procedural Control

Another characteristic of batch processes, Procedural Control is based on the Procedural Control model and is used to direct equipment-oriented actions. The Procedural Control model for these processes contains many of the coordination rules required to produce parts. The synchronization phases, messaging between transfer phases, as well as the unit tags allow the recipe to be configured in such a manner that (1) parts are not immersed in a harsh chemical if an allocated (reserved) rinsing unit is not available for the procedure, (2) robots are commanded to move

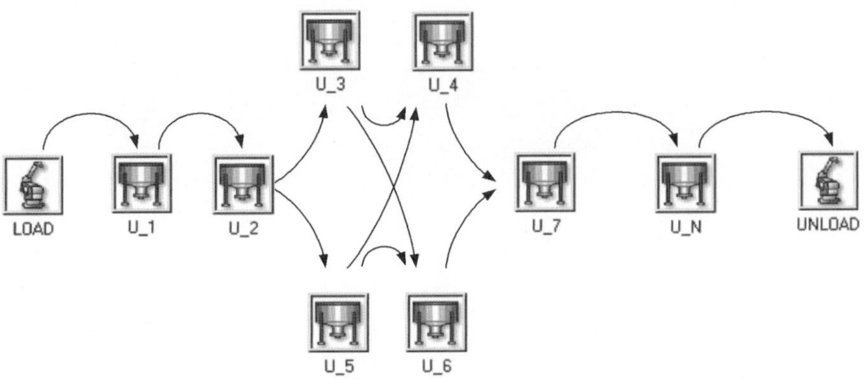

Figure 15.7. Parts processing flow sample.

to the source to pick up a part before the parts processing is complete, and (3) parts are removed from the unit as soon as the processing period is completed.

Transferring Products between Units

Transferring parts between units is accomplished using a combination of phases. The Transfer IN and Transfer OUT phases work as message partners: the X_OUT phase indicates the source of the transfer to the X_IN phase. The X_IN phase then interacts with the appropriate Robot EM. Once the part has been removed from the source, the X_OUT phase completes, and the X_IN phase continues transferring the part to the selected destination. To better understand how this transfer process occurs, the following section walks through the process of moving a part from one unit to another.

With the Transfer, Synchronize, and Pick/Place phases or basic building blocks, we can begin to specify the ISA-88.01 procedural model to perform a part transfer. Figure 15.8 contains a sample of the recipes used to transfer a part from unit 1 to unit 2. During the execution of these recipes the following steps will take place:

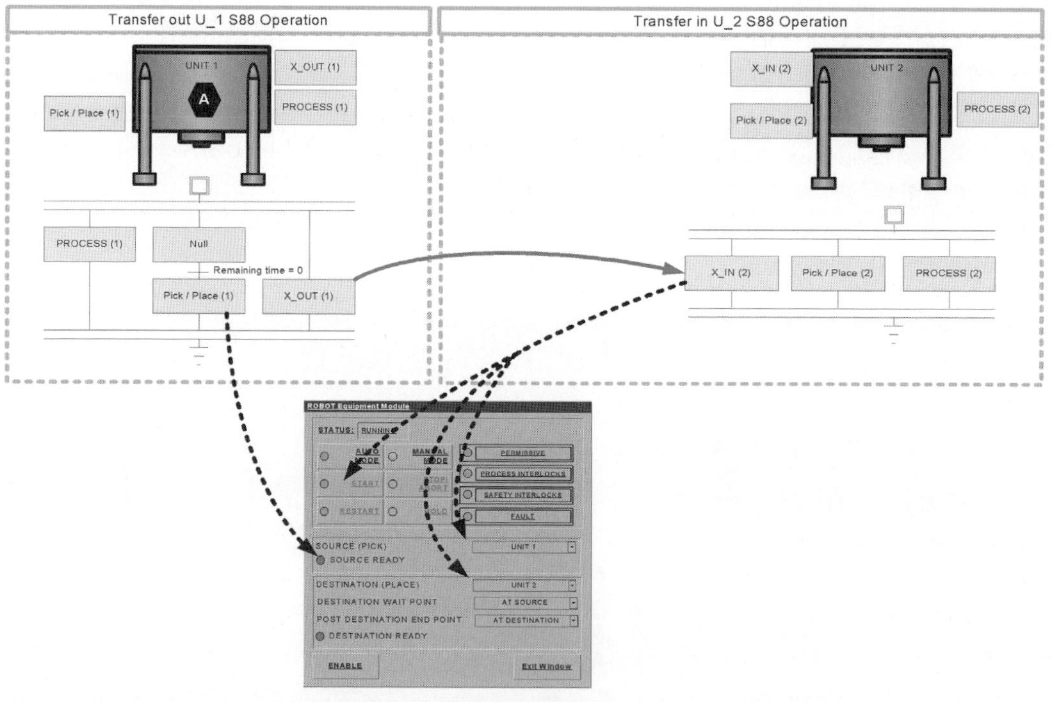

Figure 15.8.   Transfer from unit 1 to unit 2.

1. X_OUT (1) sends a message to X_IN (2). This message contains the source to move the product from.

   1.1. The X_IN (2) phase logic determines which robot will be used based on parameter selected.

   1.2 X_IN (2) makes a request to the batch server to acquire the desired Robot EM; after the EM is allocated to the phase, the logic checks for the status of the EM.

2. X_IN (2) moves the parameters to the Robot EM (source and destination information) and starts the EM.

3. The robot moves to the selected source and waits to pick up the part when the source ready flag is set.

   3.1. The Pick/Place phase sets the "Source Ready" and "Destination Ready" flags.

   3.2. X_IN (2) send a message to X_OUT (1) to complete after lifting the part and leaving unit 1. (This releases unit 1 and allows other recipes to acquire it for processing.)

4. The robot moves to unit 2 and lowers the part if the destination ready tag is set on.

5. After the part is set in place, the robot moves to the desired final location.

During this transfer it can be observed that only one ISA-88.01 procedure is running, and therefore only one batch of parts "A" is processed. Figure 15.9 shows how two ISA-88.01 procedures can be running at the same time in one physical unit (unit 2). This is accomplished using U_2 as well as its phantom unit U_2_P.

The use of the phantom unit in this case enables parts that have been processed in unit 1 to acquire a robot and start the transport process to its next destination before its destination, unit 2, has been released by the previous ISA-88.01 procedure. The implementation of these phantom units helps to reduce the overall cycle time of producing a part and also helps release the processing unit sooner. The ISA-88.01 recipe operation (class based) at a macro level consists of receiving (transferring in) the parts to be processed during the initial processing time and transferring (transferring out) the parts to the next unit during the final processing time.

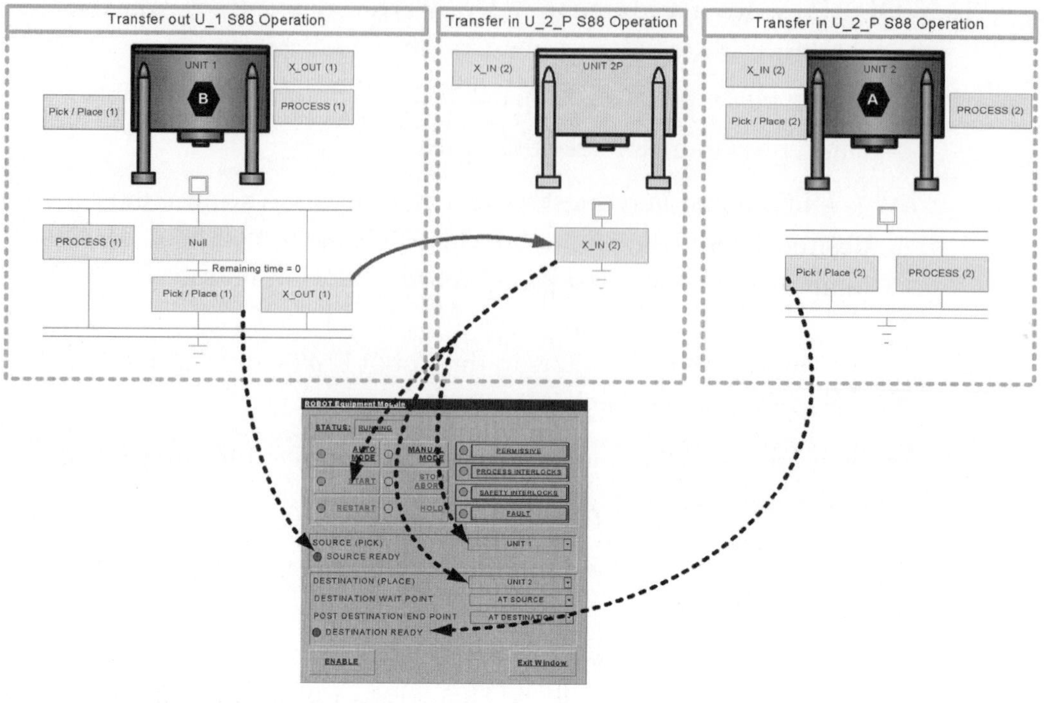

Figure 15.9.   Transfer from unit 1 to unit 2 using unit 2 phantom.

Coordination Control

Coordination Control directs, initiates, and modifies the execution of Procedural Control and the utilization of equipment entities. Coordination Control involves supervising the availability of equipment, allocating equipment to batches, and propagating mode and state changes.

The allocation of the units to be used by the procedures can be set at batch creation but should be allocated immediately before the process requires them. This allows the recipe to select a unit that has proper operating conditions. If tank or unit chemical or physical conditions are not met, the unit is disabled by automatically allocating the unit to an external user; this prohibits the unit from being used by a recipe. The robot to be used to perform a part move is requested or selected by the equipment phase. These EMs are defined in the area model as resources and therefore arbitrated as well.

The key recipe functionality consisting of equipment arbitration and allocation is partially achieved at the ISA-88.01 unit procedural level and Inter-unit Synchronization is also partially achieved at the ISA-88.01 operation level. The functionality to be achieved is as follows:

1. Only allow the parts to proceed to the next critical unit if the parts will have a safe place to go to after completing this process. (For example, only allow parts to be transferred to an acid tank if an available neutralization tank is available to transfer the parts to after this acid treatment is completed.)

2. Make the request at the latest possible time for the next set of critical as well as safe units. This allows the system to ensure that recipes bind to units that are in good operating condition. Binding the units for the entire process at the start of an ISA-88.01 procedure may cause the recipe to encounter a unit that gets put out of service after the binding takes place.

3. Allow the system or operator to disable a unit. This is achieved by acquiring the unit to be disabled via the batch server, which prevents recipes in process from using a disabled unit.

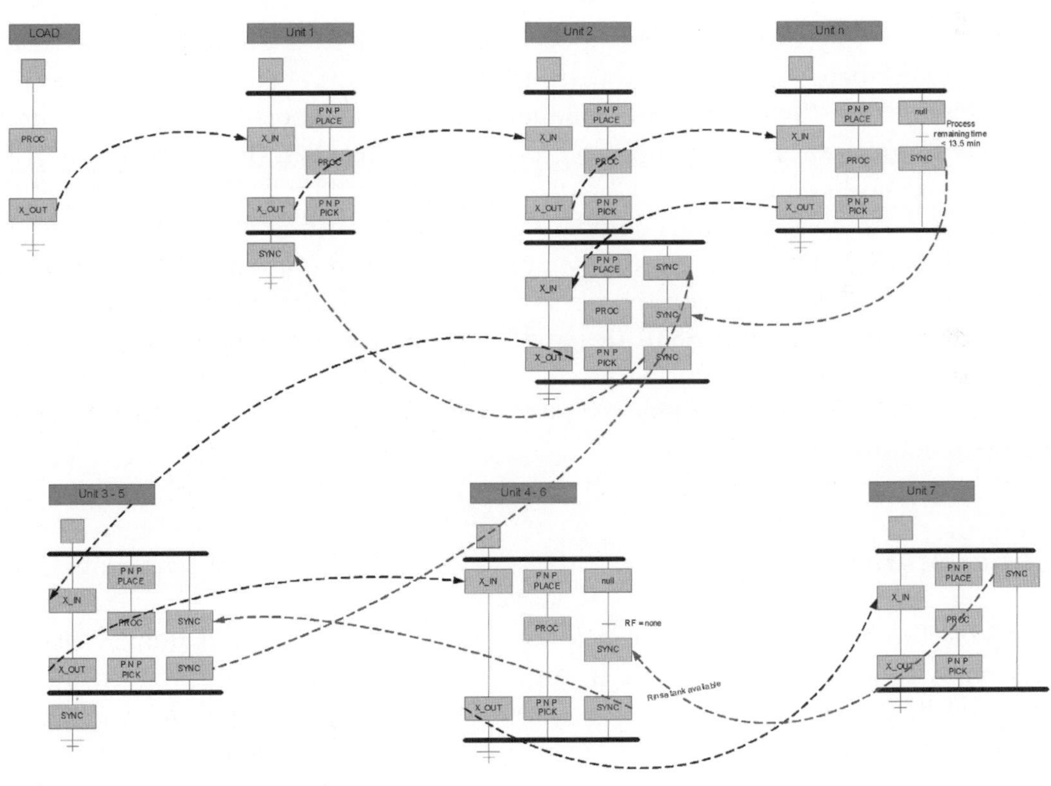

Figure 15.10. Sample operation links.

4. Use the same method for disabling units to disable the equipment that moves the parts between the units (robots/robots). This can be requested by the phase logic; if the desired moving equipment is not available, then the logic will request the next best piece of equipment to perform the task.

5. Set up the procedures to allow the batch server to select the first available units; this is a key factor for proper equipment selection.

6. Use a "pseudo" or "phantom" unit for each real unit in the process. This enables the equipment to start transferring parts from one unit to another before the receiving unit has finished processing its current part. In essence, there are two distinct control recipes using the same unit at the same time, which helps reduce the cycle time.

Benefits of Applying ISA-88.01 in Multi-station Dipping Processes

Arbitration or Coordination Control is handled through a Commercial Off-The-Shelf (COTS) batch software product rather than application-specific custom engineering code. Reporting and visualization software tools give operators and engineers significant insight into the process, enabling them to better monitor it. Flexibility in adding new parts to the process is achieved by adding new recipes to the system with no additional engineering work in Coordination Control.

An ISA-88.01-based process allows us to respond more efficiently to transfer mechanism breakdown, whether it is a robot arm or a robot moving a part from one station to another. The ISA-88.01 designed system is able to respond quickly and enables operators to issue workarounds until the main transfer system is repaired and put back in service. Reusable recipes, operations, and unit procedures help to minimize engineering.

Conclusion

Applying the ISA-88.01 design principles in nontraditional applications is beneficial to a wide range of industries. In the case of multi-station dipping processes, applying ISA-88.01 design principles provides a modular, straightforward solution to a complex problem.

Traveling with ISA-88.01: Leveraging the Standard in a Complex Assembly Process

Presented at the WBF North American Conference, Philadelphia, PA, March 24–26, 2008, by

Bruce Henne
Manager, Process Engineering
Bruce.Henne@eaglepicher.com
EaglePicher Technologies,
C and Porter Streets,
Joplin, MO 64801, USA

Marcus Tennant
Product Manager
matennant@ra.rockwell.com
Rockwell Automation,
1201 South Second Street,
Milwaukee, WI, 53204 USA

Abstract

Battery manufacturing for the aerospace and defense industries is a complex and highly regulated assembly process. By applying the models and methods of the ISA-88.01 standard, EaglePicher Technologies has enhanced assembly workflow and increased real-time process verification and quality assurance checks. These process improvements provide increased process visibility and understanding, which in turn directly decrease variability, reduce waste, and create improved process capability.

In this chapter, we will present an overview of the requirements for this complex assembly process and the concept of a traveler—traditionally a paper document that defines and documents the sequence of the manufacturing operations and the test or inspection points. Applying elements of the equipment model to the traveler concept and utilizing the procedural model have enabled EaglePicher to

1. integrate complex manual procedures with automation to achieve a flexible, accurate, and highly repeatable operation;

2. apply electronic signatures and report limit values to enable step-by-step verification of complex manual manufacturing and assembly tasks; and

3. meet stringent "track and trace" reporting requirements by leveraging event information in each subassembly and building a comprehensive complete assembled final product report that verifies all critical traveler report parameters. (This high degree of verification is essential for assembling products for critical-to-life aerospace applications.)

Industry and Requirements Overview

EaglePicher Technologies, based in Joplin, Missouri, is the leading producer of batteries and energetic devices for the defense, space, and commercial industries. EaglePicher offers a wide range of battery technology and provides other energy products and pyrotechnic devices for the defense industry, as well as advanced battery chargers and other power solutions for business, industrial, and recreational applications. The capabilities of engineers, electrochemists, and hardware assembly technicians are brought together to produce all forms of custom batteries and power supply assemblies for very demanding requirements.

Complex Assembly

The manufacturing of aerospace and defense system batteries is a very complex and intricate assembly process. Complex assembly can be categorized as a discrete manufacturing process and involves detailed assembly, interjecting active chemical formulations and micromachined engineering components. The final product is a combination of chemical and electro-mechanical engineering that achieves absolute reliability of performance in life-critical applications. Examples of other life-critical complex assembly products include life safety devices such as air bag deployment components, aircraft landing and navigation instrumentation, medical devices such as pacemakers, and other large-scale military and defense safety equipment.

Concept of the Traveler Documentation

In complex assembly, a traveler is a control plan record-keeping tracking document that is traditionally paper based. Essentially, the traveler provides and enforces a standard workflow that makes critical quality data available for the complex assembly. This record-keeping form travels with the materials as they go through intermediate operations. It is also the documentation for each of the final assembled products throughout processing and is the final record for quality.

Challenges of the Process

The requirements of the complex assembly process call for a high level of craftsmanship and demand a dexterous, thoughtful, and trained individual to perform a work task that cannot be duplicated in automation at a reasonable cost. The process requires the application of human intelligence in the assembly process and provides strong barriers to minimize variations along specific dimensions of the process. The goal is to achieve a high level of assurance that the procedure was followed exactly and was performed completely within the tolerance limits.

A standards-based method of examining the system for complex manufacturing is to consider each step in the process as a Level 1 activity in the ISA-95.01 domain hierarchy model. Each step must be clearly defined and every activity should be given a process flow and a standard procedure for the operator to follow to maximize consistency, as in an automated process.

Future system planning calls for several steps in the manufacturing process to be eventually transformed to higher levels of automation. Therefore it is a requirement to design the system with a high level of modularization sufficient to introduce automation incrementally. In addition, currently most process report values in the traveler documentation are entered manually by the operator. Continuous improvement programs such as Lean and Six Sigma require the introduction of new process entries (e.g., inputs from load cells, laboratory devices, and other analytical measuring devices) as part of the traveler record. Furthermore, customer requirements indicate a need to move reporting to electronic documentation while maintaining system flexibility for requested changes in the specification.

Complex Assembly in the Context of Physical Manufacturing Operations (PMO)

In Chapter 4, Lynn Craig describes this level of activity in a process as Physical Manufacturing Operations (PMO) (see Fig. 4.2):

> It is all about physically touching raw materials and equipment and carrying out preordained tasks. It is the focus of what seems to usually be called "operations" or "manufacturing operations" in most companies. . . . Why do we need to identify a previously ambiguous level in an enterprise? Well, while it is closely tied to Manufacturing Operations Management, it is a level that can operate on its own for significant periods of time based only on information it was provided earlier. . . . It has its own culture that has been refined and optimized over many years. More of a company's treasure is invested in and used by the operations at the PMO level than the rest of most manufacturing enterprises in total. It is a level that directly impacts not only cost but quality, capacity, reliability, yield, safety, and so on. It is a pretty important level.

In a majority of cases in manufacturing, Lynn states, "when we rely totally on manual control of procedures, there is no good reason why PMO needs distinction from other manufacturing-oriented functions. Dealing with people can be casual and need not be all that precisely organized. In a totally manual case, precision doesn't matter much. People will generally figure out what is meant and what they need to know with no harm done."

In the manufacturing environment of a complex assembly process, precision in the execution of manual procedures is a critical requirement. Additionally, there are continuing demands on the process to improve precision and accuracy in process traceability and to remove boilerplate, form-based, check-sheet documentation. These improvements are required, along with maintaining or improving manufacturing flexibility and transitioning from slow, inefficient paper-based work instructions to electronic instructions and record keeping.

Considered Approaches

Web-based work instruction tools that were evaluated did not have the enforcement capabilities necessary to track or prevent deviation by the operators in the assembly process. Many of these tools are simple checklists without the benefit of sequential-step documentation. For instance, an operator can use the tools to sign and date each step after the completion of the workflow task but not when each sequential step is performed. Throughout a workday, an assembler could deviate significantly from the desired standard operating procedure without being effectively tracked or guided to perform exact procedures.

The Manufacturing Execution System (MES) platforms currently available are designed to provide a multitude of purposes. They do not meet all the requirements desired for integration with critical sequential or real-time elements in the workflow of the complex assembly processes. Several MES packages were considered, but none of the packages fit our requirements, due to the following:

1. Their focus was on Manufacturing Level 3 activities around scheduling, materials management, quality assurance, and work cell–related workflow. They did not provide the granularity, enforcement, and exception handling that is required for Level 1 manual assembly activities such as those used in a Sequential Function Chart (SFC) method application.

2. There was no preestablished or clear pathway to migrate from manual to automation workflow. Most MES applications are designed with the manual workflow as the "End in Mind." They offer little or no integration to incorporate traditional industrial automation.

3. The event information in the MES applications examined was not detailed enough to allow in-depth process analysis and problem solving and did not meet customer track and trace requirements.

The evaluated MES applications focus on monitoring and documenting what operators do or key in rather than offering a set procedure detailing all work instructions step-by-step. They permit subtle sequence changes that are not tolerated by an SFC sequencer.

Advantages of SFC ISA-88.01 Modular Applications

ISA-88.01 SFC procedures require detailed steps to be clearly, accurately, and precisely defined, always using the same ISA-88.01 method. Driving processes this way tends to clarify "Critical-to-Quality" details and verifies that "Best Practices" are being used for the standard procedure. Therefore, as ISA-88.01 control procedures are used by the operators repetitively, they are self-reinforcing—driving understanding, confidence, and further expertise in performance of the manual operations.

How ISA-88.01 Was Applied

Combining an ISA-88.01 standard in modularization with a Commercial Off-The-Shelf (COTS) Batch package and issuing instructions to the process operators through a Web interface (rather than a controller) for detailed Level 1 work instructions permitted all the same modularization concepts to apply as they would in an automated process. This enabled EaglePicher to meet their rigorous requirements in workflow and flexibility.

Designed initially to be rolled out in a pilot facility, the solution is structured to be scaled up quickly and easily to full production to meet demands as products move out of the R&D prototype phase to full production. Some of the design details that made the solution successful were the following:

1. Designing one manual instruction step per phase was extremely helpful (Fig. 16.1). The COTS ISA-88.01 Batch software had the capability to add multiple instruction steps within a recipe phase. The complex assembly process required the ability to present one instruction step at a time to force the operator to input data, take an electronic reading, respond to any deviation requirements through applying an electronic signature, add comments, and record Corrective Action, Preventive Action (CAPA) problem report information before moving to the next instruction step.

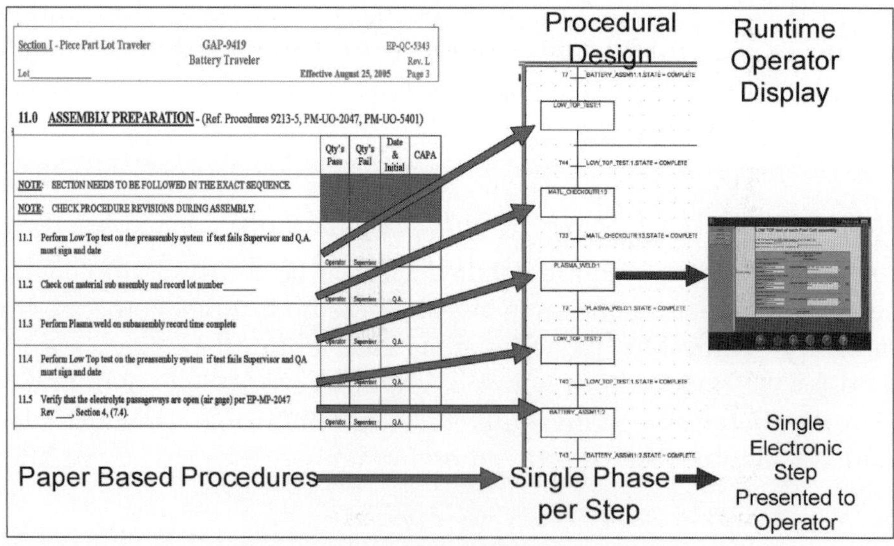

Figure 16.1.   One manual instruction step per phase.

2. Building parallel operations of phases enabled the capability to perform multipiece operations that have different timing requirements. This allowed the production system to have the flexibility to produce a subassembly, set the subassembly aside in a work-in-progress holding area, and enter additional quality assurance data collection attributes from laboratory procedures or record other electronic system inputs (Fig. 16.2).

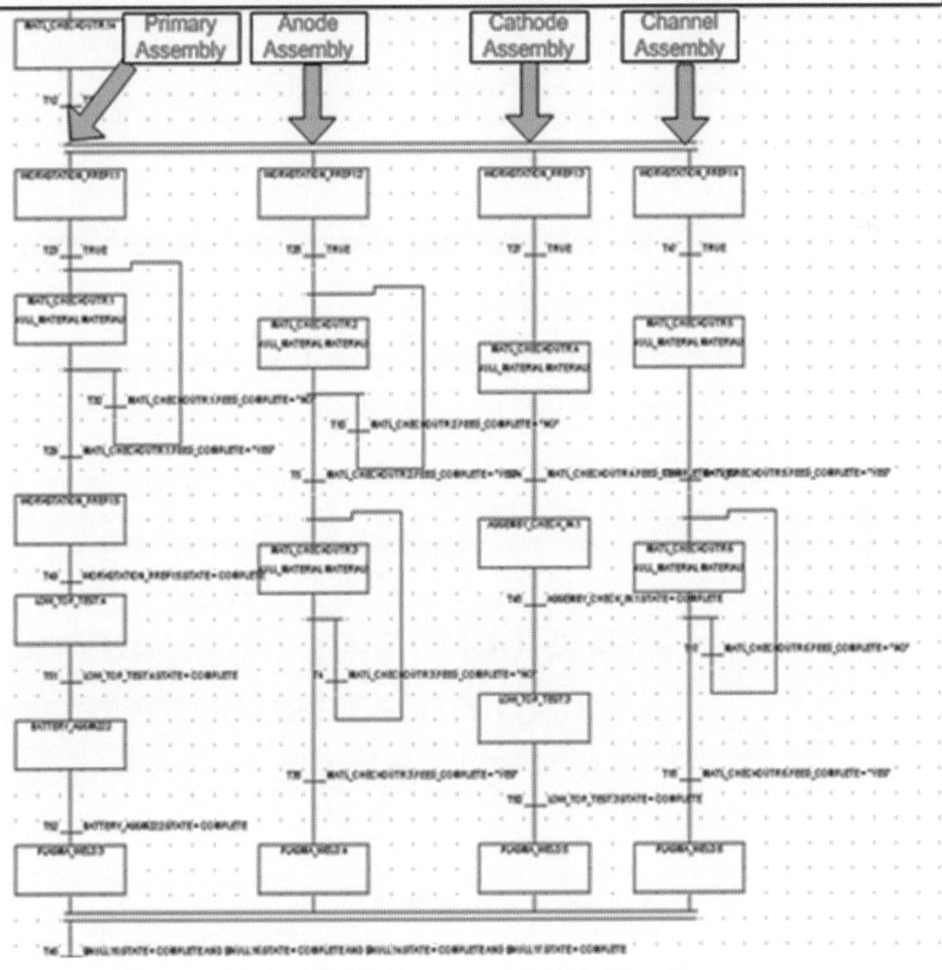

Figure 16.2.   Subassemblies built in parallel.

3. By following the ISA-88.01 model in the design of an electronic traveler, the use of unit procedures, operations, and phases for step-by-step building of subassemblies provided all the critical-to-quality control and report parameters in a hierarchical structure. The complete final product traveler is an aggregation of all the individual traveler documents that correspond to each individual subassembly. The individual subassembly travelers are useful for comparing lot-to-lot subassemblies.

4. Designing procedural steps to automate or semiautomate several manual equipment phases in the future was also extremely helpful. The primary change at the equipment-module level redirected the data source from the manual instruction data server to a controller phase data server (via Object Linking and Embedding for Process Control [OPC] or other communication protocols) with no change to the procedural or reporting structure (Fig. 16.3).

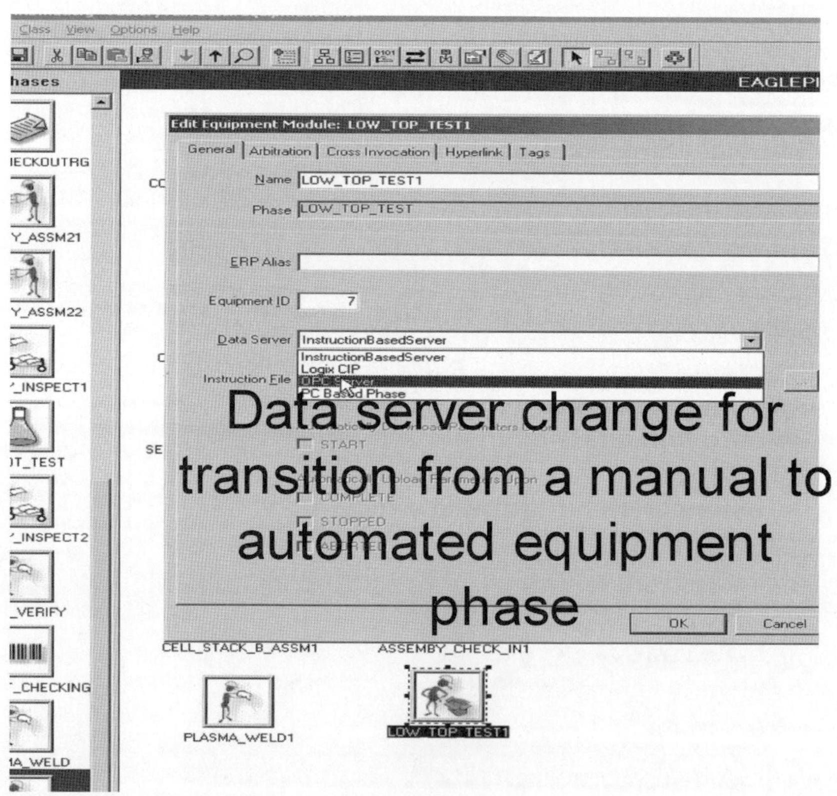

Figure 16.3.   Data server change.

Benefits

By the end of the project, several significant system improvements had been achieved:

- Use of an electronic traveler allows data to be easily analyzed using efficient database tools. New measurements are easily integrated into the electronic traveler. Systematic collection and analysis of critical-to-quality data efficiency is paramount for Statistical Process Control (SPC), Lean, or Six Sigma projects. A by-product of use is that it specifically identifies and builds knowledge about critical elements of the process. Doing this with manual data collection is either impractical or very expensive. The cost and knowledge lost in simple, manual, paper-based data collection are often not visible and therefore underestimated.

- Adherence to the ISA-88.01 methodology when designing the process enables EaglePicher to meet requirements on the standards for workflow across all quality control and manufacturing functions. This element of the design is often challenging and involves considerable discussion and negotiation, but it makes for a better process in the end.

- Effective change management is an inherently important benefit. Changes in procedures, data collection, equipment, and quality assurance checkpoints are an integral part of ISA-88.01 modular design methods.

- The rigorous enforcement of manufacturing ISA-95 Level 1 workflow forces the immediate discussion of any proposed procedure change and drives documentation of tribal knowledge from skilled operators into the standard operating procedure and electronic record.

- Use of the COTS batch system application programming interface enables a single mapping or insertion point of data exchange from the equipment layer of Level 1 and 2 to Level 3 MES and Level 4 Extraprise Resource Planning (ERP) operations.

- The significant visibility into the process increases process knowledge, which in turn is the fundamental driver for process simplification and robustness. Without this detailed knowledge, changes in process improvement can result in inefficient overengineering.

- Use of ISA-88.01 models and methodology in the design of the process from the ground up enables incremental automation to be added into the system at appropriate points.

Conclusion

ISA-88.01 modular standards applied through a batch application package successfully provided for the development of critically controlled, complex assembled products for the aerospace and defense industries. The key benefits are tight control of the production sequence, efficient tracking and analysis, and flexible and modular process control adaptability based on an international standard.

Quest for the Perfect Batch: A Batch Distillation Real Life Case

Presented at the WBF North American Conference, Baltimore, MD, April 30–May 3, 2007, by

Ellen Murphy
Process Engineer
ellen.murphy@lyondell.com
Millennium Specialty Chemicals,
a Lyondell Company, P.O. Box 389,
Jacksonville, FL 32201-0389, USA

Manuel Florez
Process Engineering Manager
manuel.florez@lyondell.com
Millennium Specialty Chemicals,
a Lyondell Company, P.O. Box 389,
Jacksonville, FL 32201-0389, USA

Abstract

Continuous process optimization in terms of product yield and throughput is an ongoing activity in every batch manufacturing facility. Increasing operating costs dictate that wasteful and inefficient operating practices be eliminated. The need for optimum performance on a batch-by-batch basis drives the effort to achieve a repetitive "perfect batch."

This chapter presents a real-life batch distillation case for which performance was driven from a state of frequent batch failures to one that has not seen a batch failure in over a year. In addition, the process performance has been improved to the point that average batch cycle times have been cut by 15% and first-time right batch yields have been increased by 208%. Process analysis methodology and actual process data are presented.

Introduction

The quest for the perfect batch prompted Millennium Specialty Chemicals (herein referred to as Millennium) to develop systems and methodology geared toward achieving such a goal. The end result can be best described as the perfect batch by design. Of course, achieving perfect batch performance on an ongoing basis is hampered by the imperfections of daily plant engineering life. Therefore the real measure of success is how close and how frequently we come to the perfect batch standard as it was designed.

Designing batch performances for processes using new equipment and advanced computer-based control technology is normally an easier task than achieving similar results with existing processes and equipment with varying levels of instrumentation and automation. This chapter presents our current approach to batch analysis, performance evaluation, and process redesign in an effort to obtain optimum and consistent results for existing processes in existing equipment.

The methodology used to achieve batch performance success was not developed exclusively for this purpose. It was, rather, one emerging from fundamental concepts, available in-house tools, and business requirements. The method is now formally known as the Process Analysis and Improvement Pathway (PAIP) at Millennium. This conceptual approach to process improvements includes methods and tools for fast and systematic process review, documentation, improvements, validation, and continuous updating. The method is flexible and highly adaptive, as it has multiple starting points and trigger conditions that provide a fluid framework for process improvements. It also incorporates many features that were specifically designed for the Millennium environment. The PAIP method is shown schematically in Figure 17.1 with explanatory notes in Tables 17.1 and 17.2.

The overall strategy and objective of the process improvements using the PAIP methodology focused on (1) removing process noise, (2) bringing the process under control, and (3) improving the process performance while keeping it under control. This concept is illustrated in Figure 17.2.

Once the process has been improved, it is critical to maintain the improved performance level. Three keys to maintaining high-level performance are

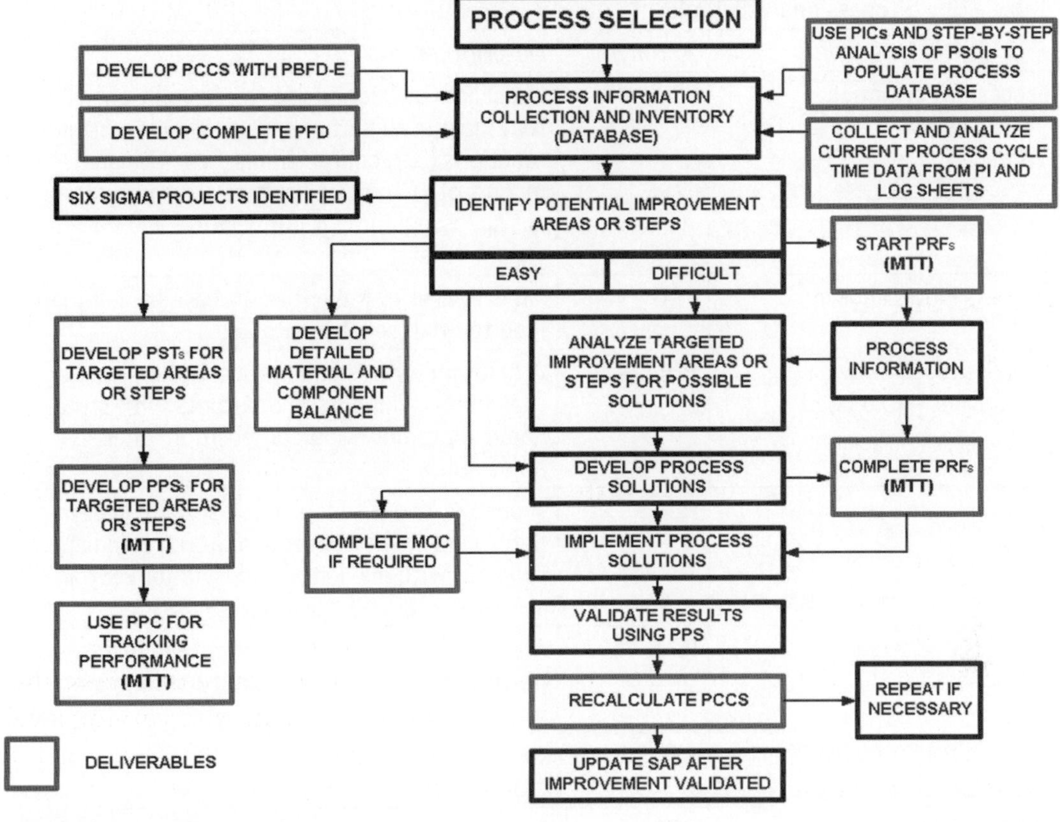

Figure 17.1. PAIP flowchart.

(1) documenting the improved process with a Process Reference File (PRF), (2) training the operators in the process and in the process performance requirements, and (3) monitoring the process by supervision to provide feedback to the operators on how well they are implementing the documented process.

Before the process improvement details are presented, the following section will describe the business manufacturing background that prompted this work, with a set of initial observations on process performance.

Real Life Case

A supply crisis with one of Millennium's higher volume products prompted the first full-scale application of the PAIP methodology. The crisis was created by a convergence of increased demand, poor reliability, high manufacturing cost, and

Table 17.1. Process analysis tools

Process analysis tools	Acronym	Description
Process Information Checklists	PIC	Checklists developed for various unit operations that include all parameters of interest, including reactions (e.g., charge times, heat transfer, mixing, etc.) and distillations (e.g., product and cut specs, equipment information and design, vacuum system design, etc.)
Process Flow Diagram	PFD	An overview of the entire process from initial feed to final (sellable) product
Process Block Flow Diagrams—Extended	PBFD—E	A diagram (where each process step is a block) that shows all inputs and outputs with stream name, mass flow, and composition where applicable
Process Cost Calculation Sheets	PCCS	A costing tool developed in a spreadsheet to determine costs from raw material to finished product, with explicit accounting for recycle streams
Process Performance Standards	PPS	Metrics developed for each process to assess performance (e.g., yield, quality, throughput, batch time, etc.)
Process Performance Charts	PPC	Charts that display trends of the performance standards
Process Sequence Tables	PST	Tables that define the process as a series of steps, lists, any inflow, outflow, cumulative mass and volume in the system, and step times
Process Reference Files	PRF	A single document that combines the other process analysis tools and serves as a compendium of information about the process
Other useful abbreviations		
Manufacturing To Target	MTT	A method of operating the manufacturing process in such a way that quality is neither too high nor too low but right on specification
Management Of Change	MOC	Documentation process required for any equipment, procedure, or chemistry changes
Systems, Applications, and Products	SAP	Accounting software

varying product quality. The product quality issues were of particular importance because in the aroma chemical business, products must not only satisfy analytical criteria but also meet a subjective odor target.

<table>
<tr><td colspan="2">Table 17.2.   Process Reference File (PRF)</td></tr>
<tr><td>PRF Sections</td><td>Descriptions</td></tr>
<tr><td>Process Description</td><td>Includes in-depth description of chemistry involved</td></tr>
<tr><td>Process Flow Diagram (PFD)</td><td>Shows all process steps from raw material to finished product, including storage for intermediates</td></tr>
<tr><td>Process Material Balance</td><td>Overall and component material balances over the process step being defined and analyzed in the PRF</td></tr>
<tr><td>Process Parameters</td><td>Anything that can be measured pertaining to the process; also, the effect of parameter deviation on process results</td></tr>
<tr><td>Process Automation and Data Historian</td><td>Defines data logging and display requirements; also defines overall control strategy and lists critical instruments</td></tr>
<tr><td>Process Performance Standards (PPS)</td><td>Metrics developed for each process to assess performance (yield, quality, throughput, batch time, etc.); includes charts to display historical trends in performance</td></tr>
<tr><td>Operating Instructions</td><td>Directions for how to make the equipment perform as required</td></tr>
<tr><td>Process Sequence Tables (PST)</td><td>Defines the process as a series of steps, lists any inflow, outflow, cumulative mass and volume in the system, and step times</td></tr>
<tr><td>Process Troubleshooting</td><td>Tips for the operator (and engineers) about common problems and solutions</td></tr>
<tr><td>Analytical Methods and Data</td><td>Lists and describes what analysis is done apart from the process and why the analysis is necessary</td></tr>
<tr><td>Computer Recipes (if applicable)</td><td>Batch sequence description, equipment module descriptions, and state tables</td></tr>
<tr><td>Waste Management and Environmental Concerns</td><td>Lists any waste streams and describes mitigation efforts</td></tr>
<tr><td>Safety Issues</td><td>Lists hazards associated with the process, required Personal Protection Equipment (PPE), and any task-specific special PPE</td></tr>
</table>

The supply crisis was of such magnitude that it required multiple resources of various disciplines and a complete immersion in the problem. A high-level analysis showed that while there were various process steps that contributed to the process unreliability, it was the last step in the process (Step 7, batch distillation of the final crude) that was proving most troublesome (Fig. 17.3). A high level of activity was directed at this process step once it had been targeted for improvement.

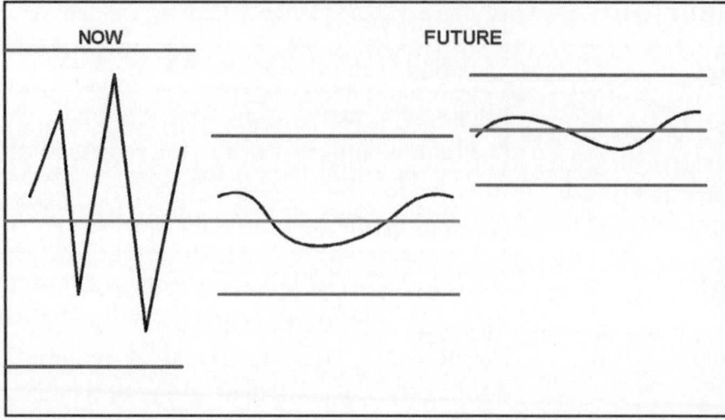

Figure 17.2. Overall strategy in using the PAIP methodology (process noise diagram).

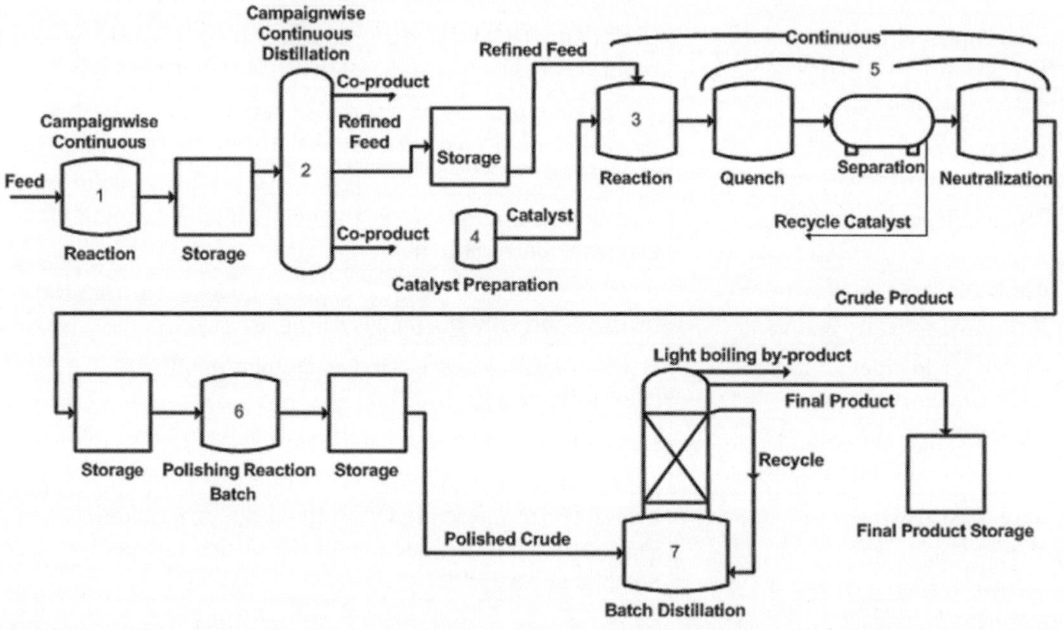

Figure 17.3. Overall process flow diagram.

The shotgun approach to troubleshooting had been tried and proven unsuccessful. Review and analysis of normally recorded data showed that the quantity and quality of the existing data was inadequate to properly define and understand the process problem, rendering any attempt to improve the process impossible. New sample points had to be installed, and the quality of the sample analysis had to be improved. The focus then turned to equipment reliability, operational issues (procedures, training, etc.), and process performance expectations. In addition, the data collection and analysis was changed from an all-paper to an electronic system to allow for remote monitoring and data analysis.

It did not come as a surprise that crude charge variability was a contributing factor for the substandard performance of the final batch distillation step (see Fig. 17.4 for data showing 5 months of poor process performance). This finding immediately spun off a second project aimed at stabilizing charge quality. At the time of this writing, five out of seven process steps have been improved and are providing incremental yield benefits.

First Observations and Fixes

An in-depth study of the sample analysis technique used for the starting pot, stream, and product tank samples showed that control room operator-generated analyses did not match the Quality Assurance Lab analyses. The Gas Chromatograph (GC)

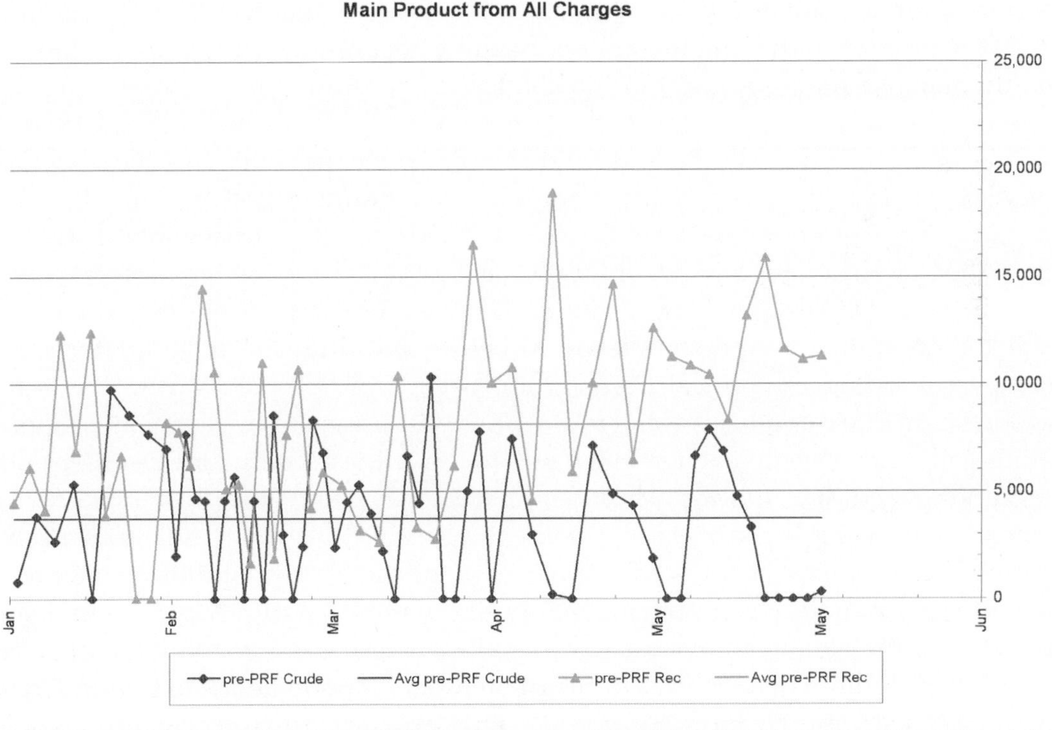

Figure 17.4.   Process performance before analysis and improvement.

methods being used were different for the pot shot, front recycle, and final product, resulting in inconsistent results in the determination of some of the more difficult to resolve peaks. The inconsistencies in the GC methods were tracked back to isolated process improvement efforts and analytical equipment changes over the years. The GC methods used were standardized so that the same technique was used for all samples. Another source of measurement error was the size of the sample shot into the instrument. If the sample size was below the sensitivity threshold value, the less abundant components were not detected. On the other hand, if the sample size was too large, close-boiling impurities did not fully separate from the main product component, which led to an underestimation of the impurity levels. After the sample analysis technique was refined, the control room analysis became more reproducible and much closer to the Quality Assurance Lab analysis. Now reproducible analytical data were available for studying process performance.

More reliable analytical techniques and increased stream sampling frequency led to the discovery that the light-boiling impurities separated during the initial lights and front recycle cuts were not reaching low levels in the distillate stream as quickly as predicted by process computer models. Because the samples were taken after the accumulator drum, at the discharge of the distillate pump (Fig. 17.5), the hypothesis was developed that the light-boiling chemical species were held in the accumulator and not effectively removed with the take-off stream.

The problem of having the light impurities linger in the distillate was caused by the relatively large holdup in the condenser, accumulator, and reflux piping. The long holdup time prevented the cut from being as distinct and sharp as it would be if the reflux stream was void of holdup components. In effect, the column external piping was providing a significant amount of back mixing, partially negating the column separation efficiency. This behavior is similar in nature to that described by Kister as "component trapping."[1]

To verify this hypothesis, a sample tap was added between the condenser and the accumulator, and samples from before and after the accumulator were compared during the critical point in the batch right before starting the main product cut. The hypothesis was proven, and work began to overcome this equipment deficiency. The first action taken was to lower the accumulator level setpoint to minimize holdup volume. Then the process was changed to lower the reflux before achieving the analytical cut point to quickly flush the distillate handling system. This solution is counterintuitive, since normally the solution to the problem of separating two components that pose a separation difficulty is to increase the reflux ratio, which allows the lighter boiling component to concentrate in the top of the column. Then, after the concentrated lighter boiler is removed, very little remains to contaminate the heavier boiler. However, for our equipment configuration, the flushing technique proved effective at removing the last tenth of a percent

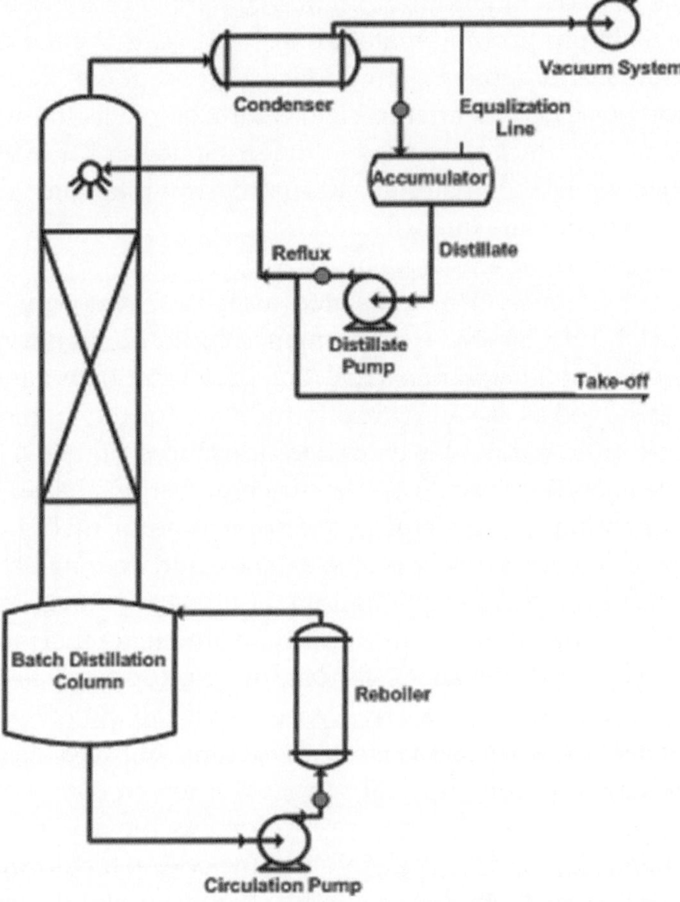

Figure 17.5.   Batch distillation column equipment diagram.

of the impurity, which would have otherwise remained in small but large enough quantities to prevent the transition to the main product cut. This reflux flushing fix allowed for some in-spec product to be generated, relieving to some degree the supply crisis. Systematic process improvements could now begin.[2]

Systematic Noise Reduction

Data: Quantity, Quality, and Odor

A temporarily higher rate of stream and pot sampling was instituted as the project moved through the initial first observations and verification of analysis phase. This higher sampling rate allowed for more exacting component balances with particular emphasis on the main product and malodorous impurities. Particular attention was paid to sample timing and frequency around fraction cut points. Accurate definition of these transition points is essential to yield optimization, as these points are normally associated with a physical change in the destination of the distillate. Yield is instantly lost when the takeoff material ends up in the wrong tank.

The increase in good data quantity and improvements in data quality allowed for statistically significant material balances. Overall material balances and batch distillation model dynamics were translated into "near real-time" balances, generated as each batch progressed so performance could be monitored "live." This capability proved very valuable in early problem detection and avoidance. These material balances served as the analytical foundation for the material balance section of the PRF document that was eventually produced (Table 17.2). Monitoring the cut sizes in real time, estimating the cut compositions based on GC stream analysis, and estimating the amount and composition of material remaining in the column helped to predict how well the batch would do long before it was near completion. Tracking composition changes in the stream samples and column temperature profiles gave the operators the ability to anticipate their next cut point.

In addition to composition specifications, the product must also meet an odor specification. After an occurrence where a product met analytical criteria but failed on odor, a "sniff GC" was run to attempt to identify the offending components. The sniff GC uses a larger than normal sample volume in conjunction with a GC method, which affords complete peak resolution. As the separated components exit the GC column one at a time, a member of the odor panel smells the emission in an effort to describe the odor characteristics of the individual components. Thus an odor profile is created. The sniff GC process identified two small impurities that boil close to the main product and imparted a "green, weedy, metallic" off note to the normally floral citrus odor of the main product. As close boilers, they were not entirely removable from the main product. However, through an odor panel trial a threshold maximum value was identified for each peak such that if the offensive peaks were below that threshold, they would not cause the product to fail on odor. A few additional peaks that were not particularly close boilers to the main product were also identified as imparting highly undesirable off odors, and this knowledge helped prevent their inclusion into the main product cut.

Equipment Reliability

After the first observations and fixes, the most urgent aspect of the process improvement effort was equipment reliability. It was imperative that all mechanical and electronic components of the batch distillation system performed as designed.

To reach these ends, extensive work was performed to improve vacuum system capacity and reliability and reboiler heat transfer performance and steam flow control. This work also ensured representative sampling in all required places. The effort was greatly rewarding and provided reliable and reproducible data that allowed the true process performance to be analyzed and compared to existing expectations. Noise external to the process was eliminated.[2]

Operational Issues

Continuing to follow the PAIP methodology, operating instructions were reviewed and batch histories were analyzed to identify additional sources of variations. The objective was to arrive at a consistent set of operating instructions that could be used to baseline the process. Operators were interviewed to extract process knowledge and compare and contrast practices between the more and less effective operators. A batch sequence table was prepared, assembling the knowledge garnered from all sources and incorporating known best practices in batch distillation. This new document served as the foundation for new operating instructions.

The operating instructions changed from a relatively vague and general document to one that incorporated all the necessary and more specific details needed for safe and efficient operation, including a section on the likely repercussions of operating errors. The concepts of reflux ratio and boil-up rates as well as the use of physical property data such as vapor pressure for water and several other key components in the feed were incorporated into the instructions. Implementation of the new operating instructions was closely supervised, and knowledge gained from direct batch observations was used to further improve the operating instructions. As the instructions were refined, operators were required to calculate the probable size of their light boilers' cuts based on the starting pot GC analysis and the mass charged to the column. The operators could then compare the predicted size of the lights cut with the totalized flow that had come off the distillation. This practice helped limit the amount of main product lost in the light boilers portion of the batch by giving the operators better feedback about where they should be at the end of the light boilers cut.

Although temperature profile changes are often used to identify cut points in batch distillation, the existing temperature and pressure instruments were not sensitive enough to detect the very small composition change that signals the difference between the front recycle and the main product. In many batches running at constant pressure, a temperature increase in the packing is the first signal of the arrival of a slightly heavier boiling component at the point of temperature measurement. However, if the separation is between two components of very similar boiling points (low relative volatility), the temperature shifts with composition are very small and difficult to detect.

The initial water removal step in the batch was streamlined as the analytical techniques were being improved. As is often the case with batch distillation charges, the crude material contained dissolved water. Although water and the lightest organic component in the process had very different boiling points and thus were easy to separate, water can still have a negative impact on column performance if it is returned to the column as part of the reflux. As heat is added initially to the stillpot, water evaporates, travels up the column, and condenses readily. While removing water, it is critical to maintain a 100% takeoff (no reflux) to prevent water from returning to the column. If water is returned to the column, it will evaporate again very quickly, causing a rapid increase in top column pressure and upsetting the mechanical (blower and liquid ring pump) vacuum system. Steam jet systems are generally more tolerant to this type of event. To prevent any return of water to the column, 100% takeoff was maintained while the top column pressure and temperature indicated that the material being removed was water. When the Top Column Temperature (TCT) began to rise, takeoff was stopped while the accumulator was allowed to build up a level of light oils. When the accumulator level reached a certain point, the entire accumulator was pumped out to the light oil recovery tank, which flushed out any water remaining in the system and lines. After verifying that the water had been removed from the system, a distillate reflux operation was initialized. This technique eliminated the occurrence of rapid swings in top column pressure during startup. Eventually the water removal technique was refined so that all water and light oils were removed with 100% takeoff until a midcolumn temperature rise was observed. This lowered the amount of operator intervention required and allowed for a continuous flush of the overhead product line, but it still meant that reflux began before the main product component was allowed to reach the top of the column.

Process Performance Expectations

Performance expectations began to be developed as each batch became more consistent through the efforts of increased data gathering and analysis, improved equipment performance, and improved operating techniques. Although at first, batches had large variations in time and amounts of material recovered in various cuts, high-performing batches were celebrated and studied so that they could be built upon. Initially, the performance target was to recover 40 to 45% of the main component as final product. This performance level is low compared to what was eventually achieved (67 to 71%). Nevertheless it was an important starting point as process consistency and reliability was established, and operators developed a sense of accomplishment.

Process Performance

Charge Composition

After the equipment and sample analysis related sources of process noise were diminished or eliminated, the focus shifted to variations in the starting pot composition. The variation had two sources: (1) the reaction train that produced the distillation feed and (2) the practice of recycling all or most of the front recycle material into the next crude batch, which meant that any variation in the previous batch performance propagated to the subsequent batch. As previously mentioned, the reaction train variability was addressed as part of a separate project. The recycling strategy was addressed as part of this effort (Fig. 17.6).

The practice of recycling material without control of amounts or composition created charge composition variations on the order of 5 to 10% for the more

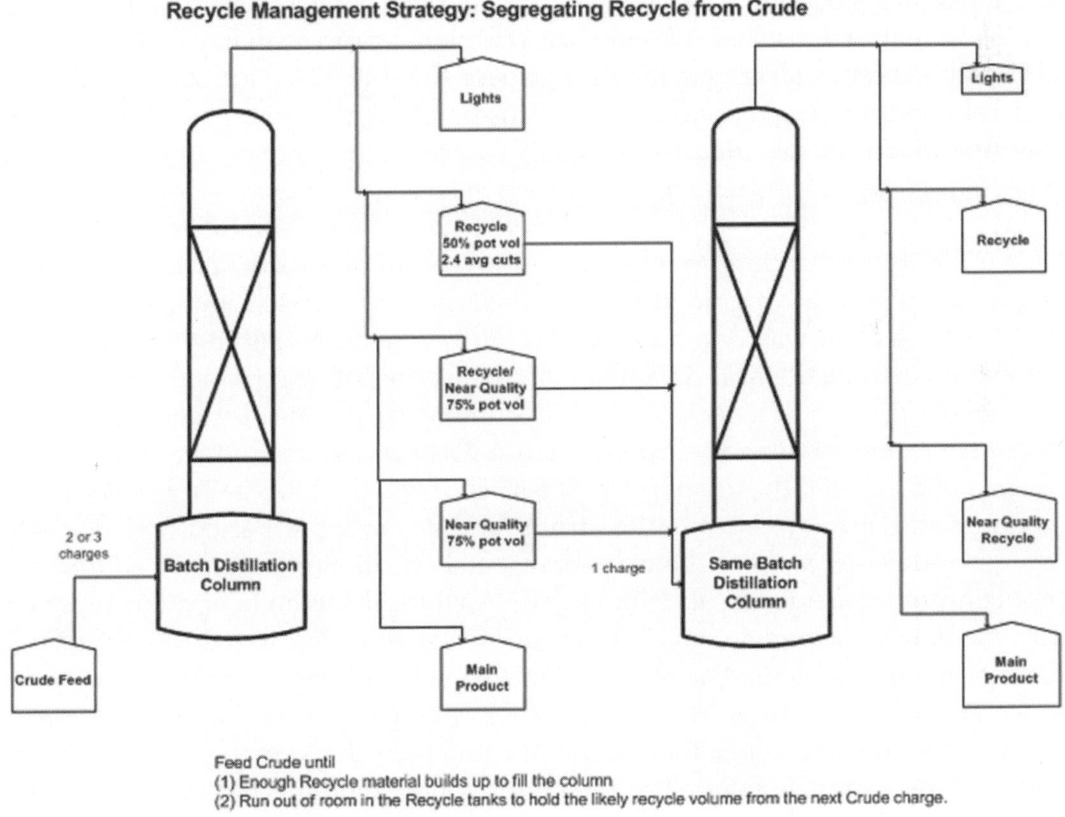

Figure 17.6.   Recycle management strategy for stabilizing charge composition.

abundant components and 200 to 500% for the less abundant components. Not coincidentally, the less abundant components—some of which are in trace or fractional percent amounts—are the same components that need regulation under both analytical and odor specifications. This was a true "recipe" for performance disaster. The practice of unregulated first-order recycle was temporarily ended. Recycle material was segregated and reprocessed separately under a different operating procedure.

PRF Development and Initial Implementation

As the supply crisis passed, much information that had been discovered, uncovered, and remembered was documented in the PRF format. The high-intensity focus on the process yielded a finely tuned set of operating instructions and a detailed set of overall material and component balances. These documents were largely ready for final documentation without any additional tweaking. Table 17.2 lists all the sections contained in a PRF along with descriptions.

The Process Sequence Table (PST) that had served as the foundation for the operating instructions was refined with additional process data including pressure and temperature ranges for various steps (Table 17.3). Process knowledge was collected and documented in the Process Description section. The Process Description section detailed the reaction process that generates the distillation charge, discussed key sources of problems in the reactions, described operation and recycle strategy of the distillation, and noted typical problematic impurities and their individual odor descriptions. A working list of all process critical variables and instruments was upgraded into the Process Parameters section. The Process Parameters section was organized based on the operator training that included the list of batch distillation parameters in their order of importance. The Process Automation and Data Historian section documented the control strategy, listed the instruments and whether their performance was process or product quality critical, and documented the use of the process historian. The Process Troubleshooting section codified the many findings from the intense focus period and includes recommended responses to typical general batch distillation problems, equipment problems, and emergency operations. The Analytical Methods and Data section documented the sample analysis techniques that were improved to understand the operations problem. The Waste Management and Environmental Concerns section documented the waste streams generated by the process and their current control mechanisms. The Safety section documented the known process hazards and their mitigation. Finally, a Process Performance Standard (PPS) was proposed, based on batch performance since the implementation of the operating instructions and the operator training sessions. The standards were discussed with

Table 17.3. Summary of results				
	Pre-PRF	*Post-PRF*	*Last half of first year after PRF*	*Second year of PRF operation*
Analysis time period (days)	150	100	182	341
Total crude batches	59	50	102	198
Percentage of crude batch failures	32.20%	8.00%	2.94%	0.00%
Main product as a percentage of crude charge	18.01%	38.40%	47.25%	55.46%
Net gain in product per batch (amount recovered for every 1 kg recovered pre-PRF)	1	1.13	1.62	2.08
Crude scaled throughput (kph for every pre-PRF kph)	1	2.57	3.25	3.92
Percentage of relative standard deviation in main product cuts (crude)	84.34%	48.10%	32.25%	12.98%
Total recycle batches	50	20	15	9
Main product as a percentage of recycle charge	39.92%	51.53%	70.32%	69.27%
Recycle scaled throughput (kph for every pre-PRF kph)	1	1.31	2.58	2.59
Percentage of relative standard deviation in main product cuts (recycle)	54.32%	37.43%	25.37%	19.25%
Overall cycle time (hours)	25	23.2	22.4	21.2

operators and production management to ensure that they were both achievable and challenging. Figures 17.7 and 17.8 compare the process performance before the process analysis and improvement, after the PRF was implemented, and the second six months of PRF implementation. The first year of PRF implementation shows dramatic average yield improvement but no dramatic improvement in process stability and reliability.

Stable Operation

Once the batches were running fairly consistently in terms of cycle times and yields, the crisis level had abated. The change that stopped the practice of first-order

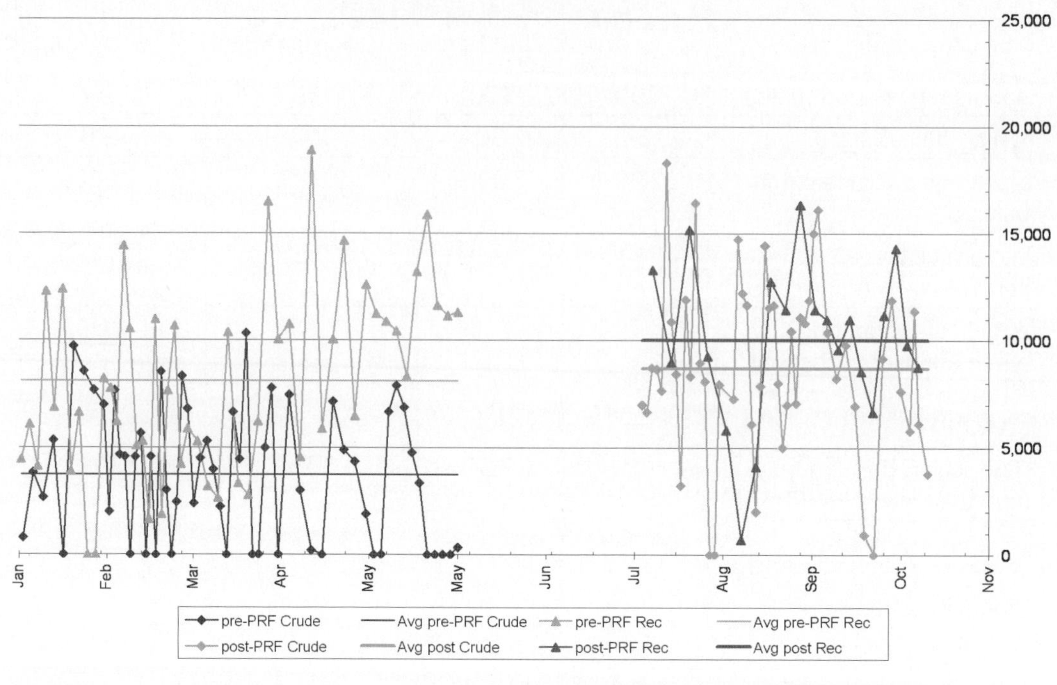

Figure 17.7.   Comparing analysis and improvement before and after institution of PRF.

recycle had made for some fairly difficult logistical problems in terms of material storage and recycle batch planning. Not enough tank space was available to continue operating in this manner, requiring storage of the front recycle cuts of several batches before enough material was accumulated to run an entire batch of recycle material. Often after several crude batches, there was not enough recycle material available to run a full recycle batch but also not quite enough room to contain the entire recycle cut from the next charge. The lack of sufficient storage space meant either that the column would be undercharged and run inefficiently or that the column charge would have to be topped off with crude material, thus lowering the main product content of the starting pot and introducing undesirable charge composition variability.

Since batches were running with some consistency and the size and composition of the front recycle cut was fairly predictable, it was time to perform some first-time-right yield optimization by manipulating the recycle cut composition to be as product rich and lights deficient as possible. With the recycle being consistently richer in the main product and slightly poorer in the light-boiling components, it was time to reinstitute first-order recycle (Fig. 17.9). Although the

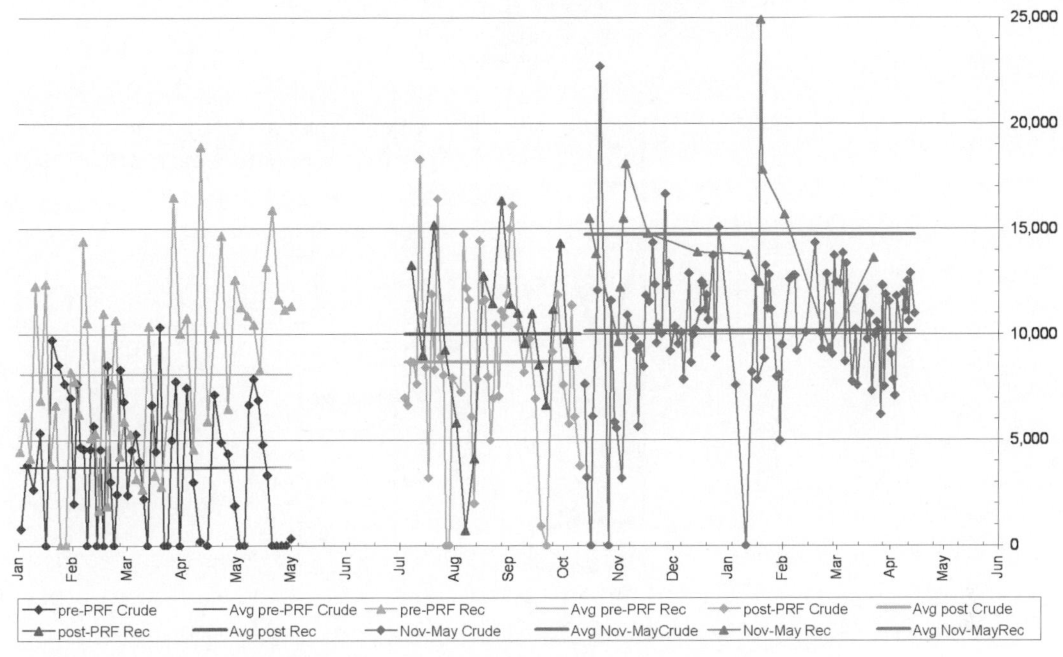

Figure 17.8.   Comparing before, after, and last half of the first year of PRF.

recycle cut volume and composition was not precisely the same every time, the variance had been greatly reduced from a few months before. Therefore the introduction of the front recycle cut to the subsequent crude charge no longer posed a risk of causing large variations in the pot composition, and it actually improved the pot composition by making it richer in the main product component, since the recycle material was biased toward the main product.

Recycle Management

Continued data analysis showed that a further refinement and optimization in recycle management could be implemented. The front recycle cut was split into two segments—the first one being the recycle generated after a top column temperature rise and the second one being much richer in main product and closer to final product quality. The high-quality recycle was blended as much as possible with the main product cut to maximize the total volume of the main product cut. The strategy of blending back was designed to maximize the benefit (i.e., larger amounts of main product generated) and minimize the risk (i.e., analytically

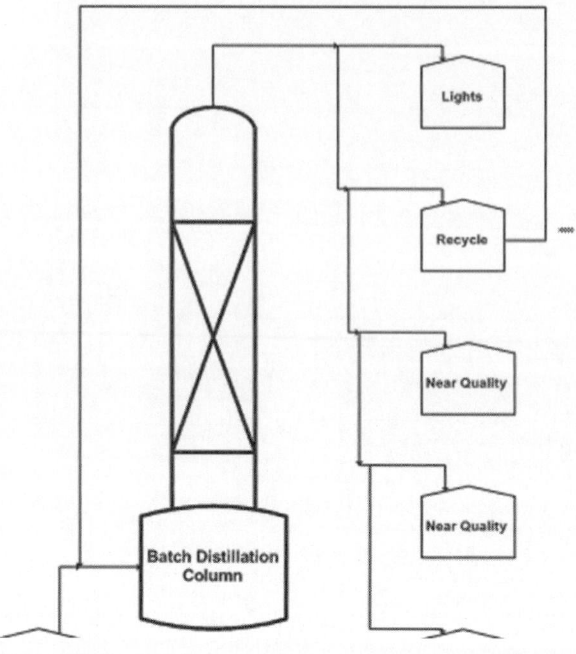

Figure 17.9.   Recycle management strategy for optimum tank usage.

off-spec or odor off-spec material). Since passing or failing the odor criteria is not entirely predictable based on GC analysis of the sample, segregating a portion of the near-good quality recycle serves to safely extend the effective cut point on many batches without risking the failure of the larger portion of the main product cut. When meeting odor criteria is not an issue, the typical strategy is to modify the analytical requirement to start the main product cut to something less than the total cut spec. Starting the main product cut before the stream sample meets the main product analytical requirements works because a later portion of above-spec material will balance out the early portion of off-spec material so that the composition of the entire cut meets specification. Figure 17.10 is a graphical representation of this. This approach of extending the main product cut afforded an average increase of 13% in product yield per batch.

Training and Monitoring

After the process was stabilized, a postprocess improvement training session was prepared for the operators. The training began with a restatement of batch

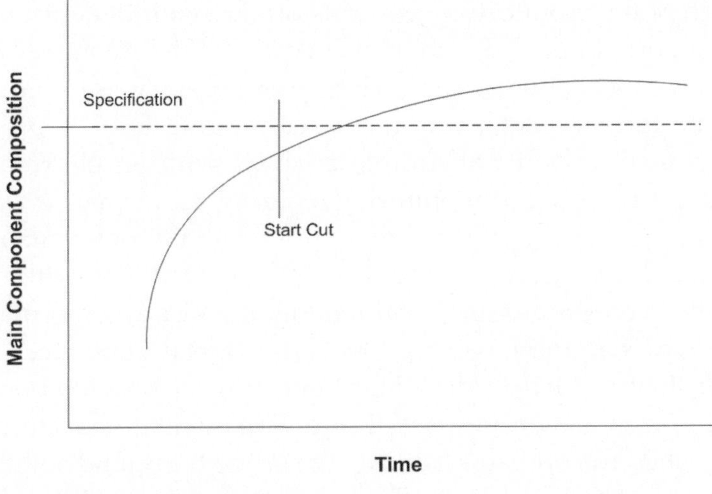

Figure 17.10.   Starting main product cut.

distillation fundamentals and discussions on process variables and their importance in process performance. The three most important variables in batch distillation are Top Column Pressure (TCP), reflux ratio, and boil-up rate. The top column pressure is the most important because it establishes the amount of heat required to achieve boil-up—the temperature at which components will boil—and can affect the relative volatility of the various components being distilled. The reflux ratio is very important because it influences the degree of fractionation of the components—how well one component is separated from another. The boil-up rate is also important because a minimum boil-up is required for the distillation equipment to be effective in contacting the liquid and vapor streams, and the boil-up must be kept below a maximum to avoid flooding the distillation column and losing efficiency. The combination of all three (absolute pressure, reflux ratio, and boil-up rate) effectively determine the column's throughput.

This training also covered some other topics of importance for process performance, such as setting the appropriate reflux ratio for different parts of the batch, using the Mid-Column Temperature (MCT) to start reflux, accurately identifying cut points, and preventing water from being refluxed. In addition, the training included a thorough review of the current operating instructions, rationale for that scheme of operation, and operational problems and solutions (troubleshooting). To maximize training effectiveness, operators should develop an understanding of batch distillation concepts like reflux ratios, boil-up rates, and relative volatility.

After the training session, continuous monitoring of distillation performance was critical to ensuring full understanding and retention of the concepts covered.

Although much of the monitoring was done off-site with data historian tools, the impact was high due to the frequency of checks and the frequency of calls to the operator on shift to ask about various data anomalies, last sample analyses, and general batch performances. Operators also communicated key process data like stream sample analysis and pot sample analysis with an electronic data sheet based in Excel. At the start of monitoring, column performance was checked at least once a day, first thing in the morning, but generally performance was monitored throughout the workday. Through daily conversations with operators, the training concepts were emphasized and reiterated. Occasionally small errors were caught before they became larger problems, but mostly operators were already aware of what they needed to do. The phone contact provided encouragement for obvious good decisions, reinforced the general distillation concepts covered in training, established the operator's performance visibility, and communicated the importance of the operator's role. As the opportunities for positive feedback arose, the operators were commended for their consistently high-level performances. Encouragement was important to maintain morale and focus on the problem as a surmountable challenge on which progress was being made. The consistent monitoring served to ensure that previously observed problems had been solved and that process noise was reduced, to identify opportunities for process optimization, and to drive the overall process toward the "perfect batch by design" goal.

One of the previously observed problems was that, especially at higher reflux ratios, a light impurity would decrease to a certain point and then get no lower. Based on the fairly large accumulator holdup and the comparison of condensate samples and reflux samples, it was discovered that this impurity accumulated at low levels in the distillate handling piping and was not flowing out with the take-off. Since the operators entered their stream sample analysis in an electronic data sheet, the monitoring included leading questions like "Your light impurity seems to have stopped decreasing over the last few samples. Do you think it is time to decrease the reflux ratio and flush out the accumulator?"

The monitoring helped reduce process noise, especially noise generated by variances in operating techniques. Typically a mistake or oversight would be caught based on process historian data, and the oversight could be corrected before it caused a significant process variance.

Consistent monitoring served to develop opportunities for process optimization in several different ways. The phone calls with the operators were a forum for them to communicate opportunities that they saw for improvement, from techniques to speed startup, and allowed them to request tools to improve their decision-making capability. The constant focus on the process led to new insights in operating techniques and new ideas for equipment configuration to improve performance.

As the small process improvements compounded and process errors were minimized and used as teaching opportunities, the process came closer to the ultimate product recovery as determined by computer modeling results. The constant monitoring and reporting of results ensured that the gains would accumulate, stay as current process knowledge, and not become lost and forgotten or overwhelmed by quotidian concerns. Figure 17.11 shows the impact of consistent monitoring over two years of PRF implementation, which caused the process performance to improve in terms of both yield and consistency.

Process Optimization

Performance expectations were augmented as batches became more consistent through increased data gathering and analysis, improved equipment performance, and improved operating techniques. The sequence table was streamlined and finely tuned. Each process step was studied over past batches, and the amount of time and acceptable variation for each step was determined based on statistical analysis of past performance. Choosing the step time around the seventy-fifth percentile of past

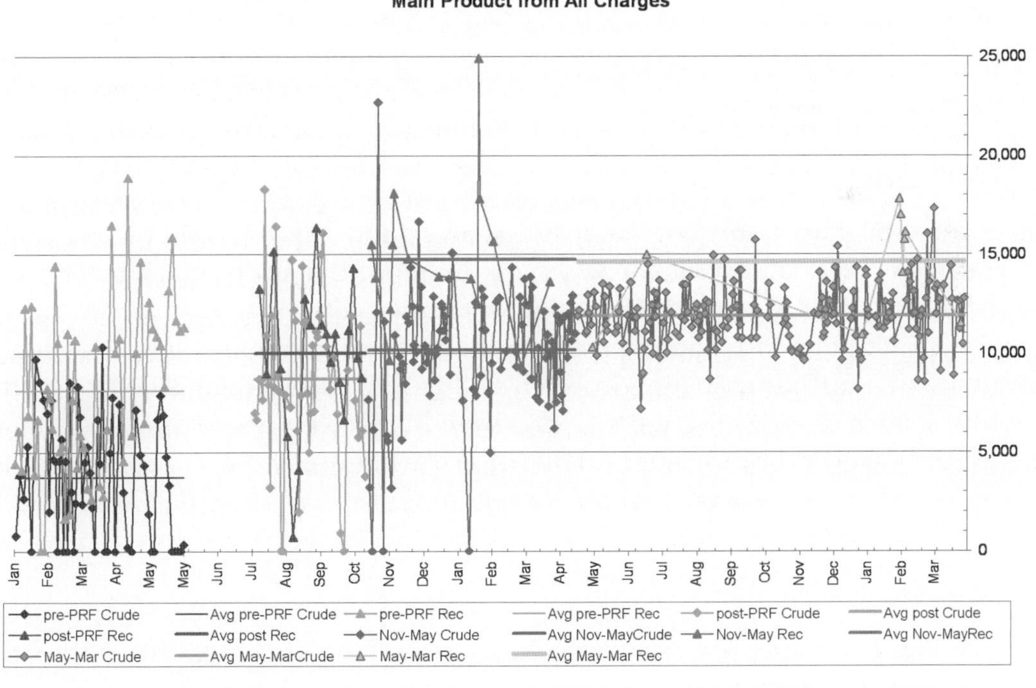

Figure 17.11.   Comparing before, after, last half of first year, and second year of PRF.

performance ensured that the performance expectation would be realistic while still challenging operators to improve overall performance. The performance expectation for the sizes of the various cuts was developed in a similar fashion, looking at past performances and variations. Although the ultimate goal was to achieve the maximum possible model-based recovery of the main product component charged in the final product cut, setting the performance expectation too high could make its achievement seem impossible. Achieving consistent performance recovering 50% of the main component as final product on every batch was important to do first and then we could raise expectations and build on past accomplishments.

A sequence table (Table 17.4) describing the "perfect batch" was assembled based on the best actual performance of each step in terms of time and product generated, minimizing the size of recycle cuts and maximizing the size of the final product cut. By separating process steps and honing each step to maximum efficiency and maximum effectiveness, the "perfect batch" can be designed. While working to achieve perfection, it is important to remember the effects of earlier process steps on subsequent process steps. Sometimes extra time in an earlier process step must be traded to reap larger yield benefits in later portions of the batch. This is especially true for using higher reflux ratios during the front recycle segment, which makes the step take longer but ultimately yields a smaller amount of recycle and leaves more main product behind for the heart cut.

Summary of Results

Batch Failures

During the 150 days before beginning our analysis and improvement program, the process step had a failure rate of 32% (very little or no main product generated). In the first 100 days after the PRF was put into practice, only 8% of batches failed. After instituting the PRF and intensifying the monitoring regime, the following 6 months saw a further reduction in batch failure rate to less than 3%. The failure rate was driven to zero after the PRF had been in place for a year (Table 17.3). By this time, however, improvement work on earlier process stages had started to take effect, blurring the analysis boundaries.

Batch Cycle Times

The pre-PRF batch cycle times averaged 25 hours. The cycle times during the PRF development phase averaged 23.2 hours. The first 6 months of PRF monitoring showed a further decrease in batch cycle times to 22.4 hours. The last 2 recorded periods demonstrated average cycle times of 21.2 hours. Overall a total decrease of 15% in batch cycle times resulted (see Table 17.3). Figure 17.12 shows the trend in average batch cycle times over 5 periods, from before implementation of PAIP to the second year of the implementation of the PRF. Figure 17.13 shows batch cycle time distributions for the same five periods. Achieving an average design cycle time of 19.75 hours is still our goal.

Batch Yields

A progression similar to that seen in the reduction of batch cycle times was seen with increased yields. On a 1000-kg batch basis, the first-time-right yields jumped by 113%, 162%, and 207%. The ultimate yield improvement target is 263%, based on a target main product recovery of 85%.

Batch Overall Throughputs

The combined effect of increased yields per batch and reduction in cycle times afforded a 245% improvement in overall process throughput. This is a truly remarkable result.

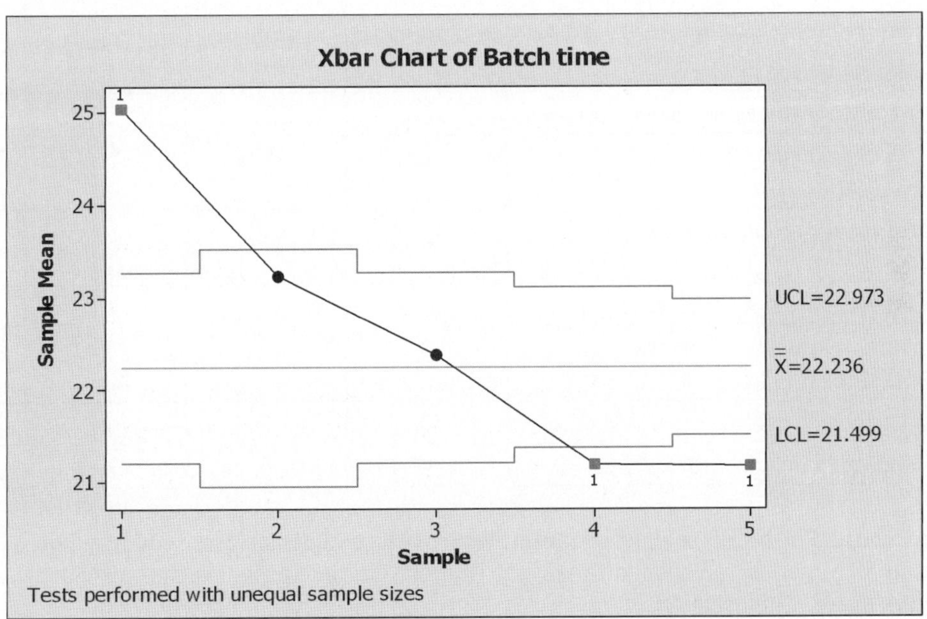

Sample 1 = Pre-PRF, Sample 2 = PRF development, Sample 3 = First six months of PRF implementation, Sample 4 = Second six months of PRF implementation, Sample 5 = Second year of PRF implementation

Figure 17.12. Average batch time changes resulting from PRF development.

Table 17.4. Batch distillation sequence table

Sequence table for batch distillation of crude "product"

Process type (batch distillation)
Equipment name

Step number	Step description	Weight (kg)	Cumulative weight (kg)	Head temp. (°C)	Stillpot temp. (°C)
1	Prepare still	0	0	< T1	< T1
2	Charge crude plus recycle from last batch	10,000	10,000	< T1	T2 to T3
3	Pull vacuum for dewatering	0	10,000	T2 to T3	T4 to T5
4	Start heating	0	10,000	T4 to T5	> T6
5	Heat up column; take off water; no reflux	−93.5	9907	T7 to T8	T9 to T10
6	Ramp pressure to midrange target over 45 min.	0	9907	T11 to T12	T13 to T14
7	Take off light-boiler cut; no reflux until MCT rise, then R1:R2 reflux ratio	−1682	8224	T11 to T12	T13 to T14
8	When TCT = X1°C and TCP ~X2 mmHg, take off front recycle	−2009	6215	T15	T16 to T17
9	Ramp TCP to lowest target over 90 min.	0	6215	> T18	T16 to T17
10	Take off near quality front recycle	−1121	5093	T19 to T20	T21 to T22
11	Take off main product	−4673	421	T19 to T20	T21 to T22
12	Take off back recycle (if necessary)	0	421	T20	> T23
13	Reboiler circ "smart" pump stops or take-off stops	0	421	> T24	> T23
14	Break vacuum—cool	0	421	–	< T1
15	Pump pot residue	−421	0	–	< T1

Notes:

A Check valve lineups. Flush LRP sealant decanter. Verify reflux valve is closed. Analyze pot.

B Steps 2, 3, and 4 happen virtually simultaneously. Cumulative time assumes that all charge activities and pulling vacuum occurs simultaneously.

C Accumulator level develops before takeoff of water.

D Steps 5, 6, and 7 happen at the same time.

E Start ramp ~1 hour after front cycle cut starts.

F Assume pot residue was balance of mass and volume. Note also that residue is not pumped on every batch

Product: "Main product"
Max volume allowed: XXX gal (10,000 kg basis)

Pressure (mmHg)	Density (kg/gal)	Volume (gal)	Cumulative volume (gal)	Step time (min)	Cumulative time (min)	Notes	Design basis
P1 to P2	–	0	0	5	5	A	–
P1 to P2	D1	3186	60	65	B	Charge pump capacity	–
P3	–	0	3186	90	95	B	Vacuum system normal capacity
P3	–	0	3186	15	95	B	–
P3	D2	25	3161	75	170	C, D	Heat transfer calculations
< P3, > P4	–	0	3161	45	170	D	–
P5 to P6	D3	560	2601	180	275	D	Model prediction for maximum flow at startup
P5 to P6	D4	640	1961	255	530	–	Model prediction for optimum RR
< P6 to P7	–	0	1961	0	530	E	–
P7	D5	356	1604	140	670	–	Model prediction for optimum RR
P7	D5	1336	269	430	1100	–	Model prediction for optimum RR
< P7	D6	0	269	0	1100	–	–
< P7	–	0	269	15	1115	–	–
P1	–	0	269	25	1140	–	–
P1 to P2	D7	260	9	45	1185	F	–

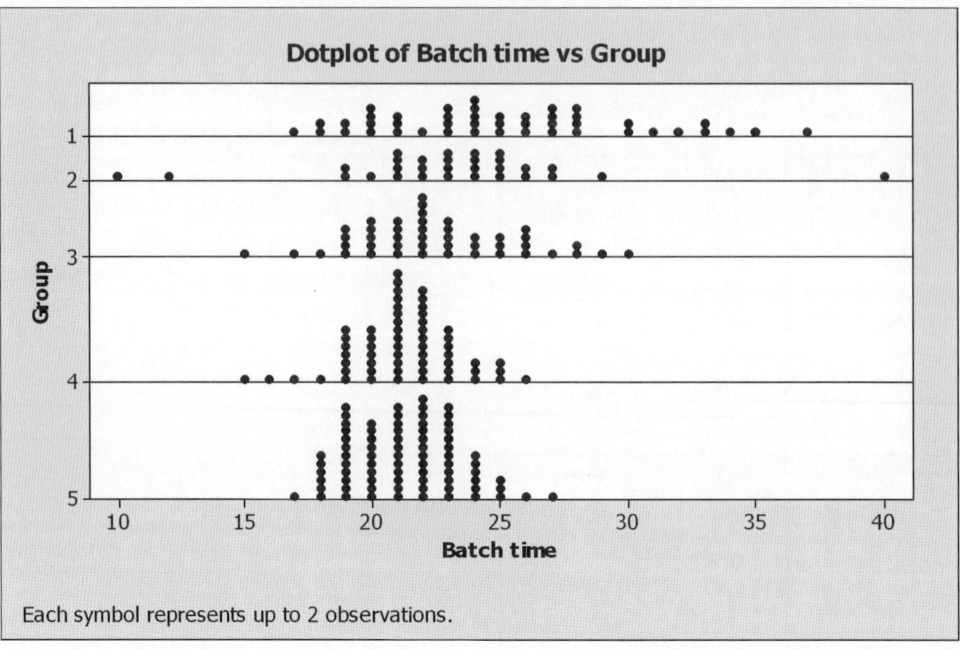

Group 1 = Pre-PRF, Group 2 = PRF development, Group 3 = First six months of PRF implementation, Group 4 = Second six months of PRF implementation, Group 5 = Second year of PRF implementation

Figure 17.13.   All batch times: Changes resulting from PRF development.

Closing Comments

The technical improvements achieved during the execution of this project allowed Millennium to continue to reliably serve the demands of its customers and develop the potential for increased sales and profitability. The increased sales and profitability are based on the improved yields and cycle times that have resulted in an improved cost structure and newly created capacity.

The efficacy of the PAIP methodology was demonstrated for the Millennium manufacturing and business environment. The approach, however, can easily be adapted to other business environments and manufacturing facilities since it is based on scientific and engineering fundamentals and best practices. The use of the methodology has been extended to include batch and continuous reactions and continuous distillation projects. Perfect batch results are not yet achieved on a regular basis, but the quest continues.

Acknowledgments

We would like to thank Kimm Antic, CI Plant Manager; Christi Leslie, Area Manager; and the area operators Linton Henderson, Chris Bray, Joel Wilcox, Howard Buie, Alex Kinlaw, and Don Minix for their hard work, dedication, feedback, attention to detail, support, and other things too numerous to mention.

References

1. Kister, Henry. 2004. Component trapping in distillation towers: Causes, symptoms and cures. *CEP Magazine*, August 2004: 22–33.

2. Florez, Manuel R. Batch distillation: Practical aspects of design and control. WBF North American Conference, April 2002.

Unit Shutdown Design for a Batch Control Application

Presented at the WBF
North American Conference,
Baltimore, MD,
April 30–May 3, 2007, by

Shashank Paithankar
Shashank.Paithankar@honeywell.com
Honeywell Automation India, Ltd.,
56 and 57 Hadapsar Industrial Estate,
Pune 411013, India

Abstract

It is my experience that the design and implementation of shutdown logic in a batch control application is often done in the Hold and Abort handlers of unit phases. This is evident because most operating characteristics and interactions between unit equipment are observed and controlled by unit phases. But this approach requires a significant amount of programming for Hold and Abort handlers—most of which can be identified as redundant.

The general purpose unit shutdown strategy discussed in this chapter will demonstrate the benefits of distributed shutdown handling. The following are the four main elements of a generic unit shutdown strategy:

- Equipment Module (EM)
- Unit Shutdown Control Module (USCM)
- Unit phase
- Procedural recipe

A USCM using sequential control functions of a Distributed Control System (DCS) has already been implemented at the facility of a large pharmaceutical company in Puerto Rico. A unit shutdown strategy integrating these four shutdown strategy elements results in increased reliability and code optimization, solving control logic deadlocks for shared equipment. The batch reporting function requirements for this strategy are discussed in this chapter. An example implementation using a DCS is depicted at the end of the chapter.

Introduction

A typical batch control system is capable of handling the discrete control, regulatory control, and sequence control handshake between different processing units through the use of the ISA-88.01 Physical and Procedural Model. Exception handling of the batch control system is crucial to avoid any undesirable effects on the safety, integrity, strength, purity, and quality of the product. Design and implementation of exception handling are reported to constitute 40 to 50% of batch design and implementation effort.

This chapter details a unit shutdown strategy for handling exceptions in the typical ISA-88.01 implementation. Unit shutdown for a batch control application uses an EM and a USCM for the implementation of exception handling for unit-owned equipment. The unit phase has to handle shutdown for acquired shared equipment. In the event of batch server failure, increased reliability is provided by empowering the USCM to identify shared equipment ownership and control.

Elements of Unit Shutdown

Arbitration

In most batch control systems, equipment arbitration is handled by the batch server. A generic design approach for custom shared equipment handling is used to facilitate arbitration in the absence or failure of the batch server. A string parameter "Owner" is used in all CMs, EMs, and USCMs. If a CM is acquired by an EM, then the equipment ID of the EM is written to the CM. Similar situations apply for unit phases acquiring EMs, CMs, or USCMs. Before the resource is released by a higher level entity, the "Owner" of the resource is set to blank. This arbitration strategy is especially useful for handling utility resources, wherein unit acquisition by a procedural recipe is not possible because of the desire to optimize shared resources.

Equipment Module (EM)

An EM can be state driven or equipment driven. State-driven EM architecture is depicted in Figure 18.1. A typical EM is implemented using a sequential control function in the DCS. This makes implementing the state transition diagram easier for different state handlers. The state transition diagram is depicted in Figure 18.2.

An EM acquires associated CMs and retains them during Hold and Stop logic execution. An equipment CM has failure monitoring configured within it. The EM looks for failure in acquired CMs, and process critical conditions are configured. One of these failure monitor conditions is a failure triggered by the unit phase or USCM.

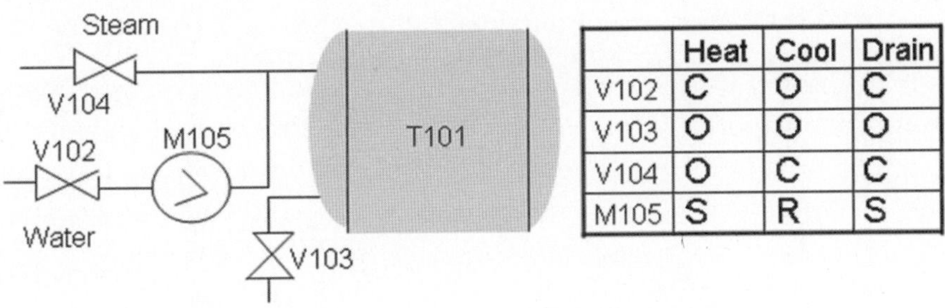

	Heat	Cool	Drain
V102	C	O	C
V103	O	O	O
V104	O	C	C
M105	S	R	S

Figure 18.1.   EM concept.

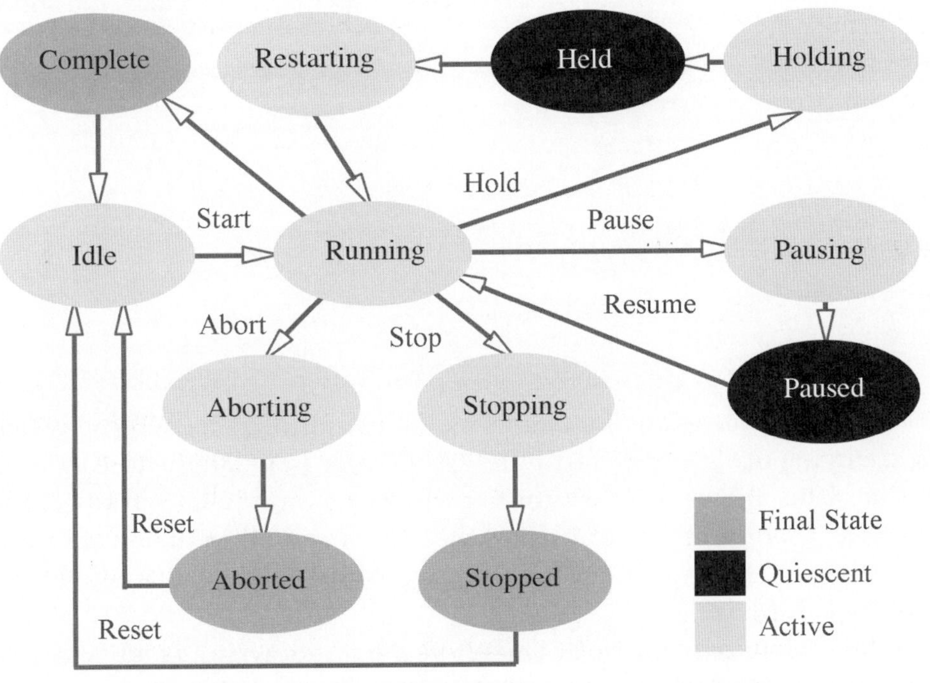

Figure 18.2.   ISA-88.01 state transition diagram.

Exception Handling in the EM

When Failure is triggered, Hold logic of the EM drives acquired resources to a Fail Safe state. Once in the Held state, the EM allows operators to manipulate individual CMs by setting the CM mode attribute to Operator Mode. This approach allows failure monitoring and shutdown implementation under a single equipment header, rather than implementing them in the Hold and Abort sequence of individual unit phases manipulating equipment in Run logic. A User Requirement Specification should have a table indicating fail-safe positions of all equipment involved. Exception handling through unit phases should handle variation from this table only.

Failure from the EM is propagated to the unit phase. Aborting the unit phase causes an EM to release acquired CMs.

USCM

The USCM is implemented using sequential control function in the DCS. The USCM is designed to provide (1) Unit State Control, where alarms may be enabled or disabled based on unit state or equipment process conditions, and (2) Unit Shutdown Monitoring and Control, where modules owned by the unit are driven to a safe state when a shutdown condition occurs or an operator selects the shutdown button on the unit display.

The following states are implemented in a USCM:

- Idle

- Operate

- Clean In Place (CIP)

- Maintenance

- Shutdown

The USCM acquires and maintains the list of unit-owned equipment. It allows for the changing of CM mode attributes to operator mode in the Idle and Maintenance state. This allows the operator to manipulate individual sets of equipment. When active, every unit phase acquires the USCM and other shared equipment in the setup step. At the end of the Run state and in the Abort state, the unit phase releases the USCM.

A Failure Monitor in the USCM monitors the shutdown initiated by an operator from the Human Machine Interface (HMI) display or the unit recipe from the Batch Operator Interface. The following conditions are also monitored by the USCM:

- Process Critical Unit Alarms

- Failure of owned or acquired equipment.

- Good Manufacturing Practice (GMP) Alarms

- Building Management System Alarms

When a unit shutdown condition is triggered, the corresponding USCM executes a shutdown sequence. The shutdown handler sequence consolidates logic to identify active unit phases and acquired shared equipment. Unit shutdown is propagated to the acquiring recipe. The operator can command shutdown in abnormal situations by pressing a shutdown button on the HMI display.

Pharmaceutical batch control applications usually require exception handling to be captured in the Batch Report or Batch Event Journal. This can be achieved by designing a USCM to run as a manual unit phase with execution taking place in the DCS controller. Procedural recipe control puts scan time constraints on the USCM, but asynchronous execution at faster scan times in the DCS controller is always possible. In the event of a batch server failure or loss of synchronization with the procedural recipe, a USCM is capable of handling a safe unit shutdown.

Unit Phase

A unit phase is implemented using the ISA-88.01 example state transition diagram depicted in Figure 18.2. The Running state logic is where the normal logic is defined to perform the actions necessary to achieve the purpose of the phase (e.g., Fill, Agitate, Diafiltrate, Transfer). The Running state logic must place all modules that are owned by the phase into Program mode so they are not accessible to the operator while the phase is running. Once acquired, shared resources are placed in Program mode. Exceptions to this rule may be required for cases where the operator needs to maintain accessibility to a certain module.

Unit phase Hold and Abort logic is designed for handling shared equipment shutdown. The unit phase monitors the phase-specific failure conditions. If shutdown is triggered while the phase is running, the Hold sequence drives shared equipment to a fail safe state and triggers shutdown in the USCM. The USCM then commands unit-owned EMs and CMs to a fail safe state. Decision logic for immediate execution and delayed execution would be configured as directed in a user requirement specification. The EM handles shutdown for the devices it owns. Figure 18.3 depicts the unit shutdown strategy.

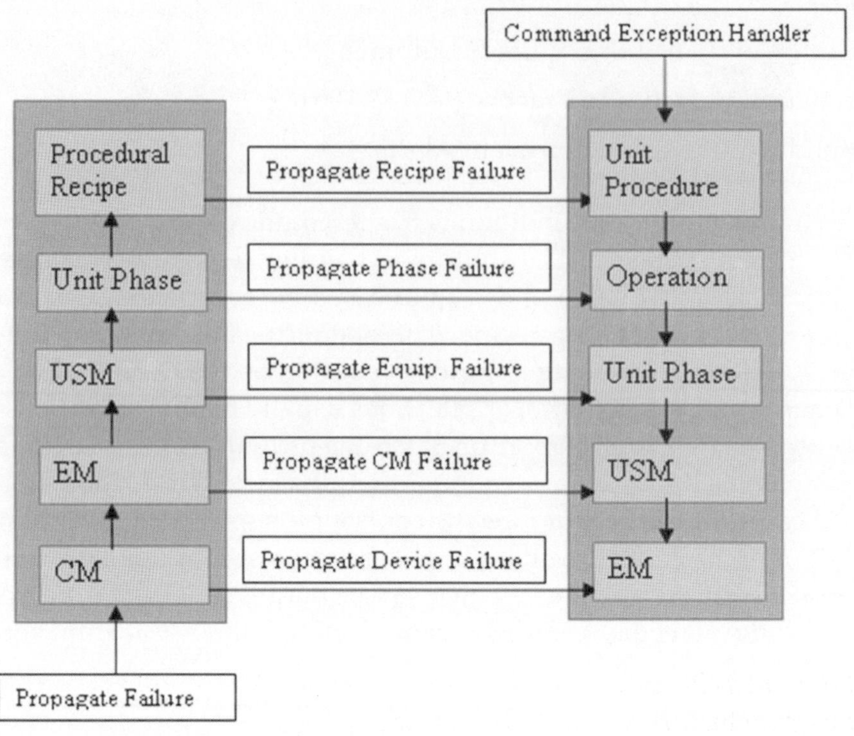

Figure 18.3.   Unit shutdown exception handling.

Phase Restart Pointers in Run Logic

The intent of using a restart pointer in the Running state logic is to branch around steps that have completed and should not be repeated on a restart from the Held state. While some configuration engineers may prefer to minimize the total number of steps in a phase by placing conditions or delaying actions based on the restart pointer value, this complicates the code and makes troubleshooting difficult. It is recommended practice to use restart pointers that check conditions after the setup logic of every phase.

Unit Shutdown Design Using Controller-based Unit Operations

Figure 18.4 illustrates the unit shutdown strategy using a DCS. Minimum system requirements to support unit shutdown strategy are as follows:

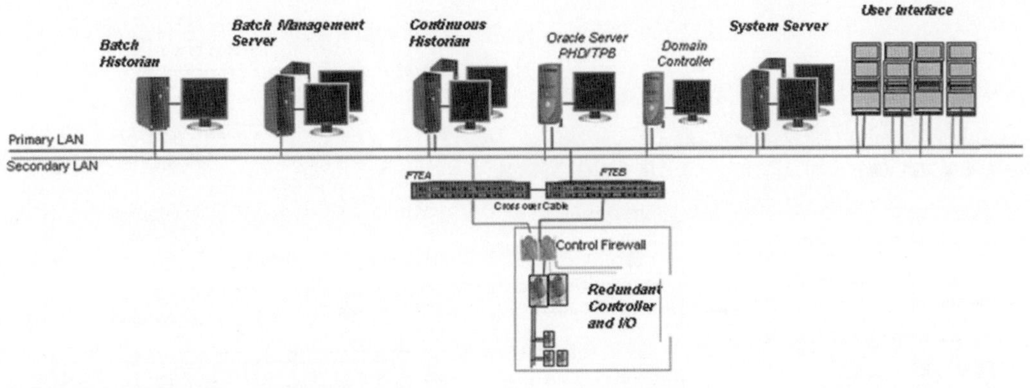

Figure 18.4. DCS system architecture.

- *DCS server.* This is a computer on which DCS database software and utilities run. A DCS server can offer client-server architecture or a peer-to-peer interface for HMI operator stations with DCS controllers, depending on the configuration.

- *Continuous historian.* This serves a data collection to provide continuous data history for batch reporting.

- *Procedure analysis.* This provides the batch report and history functions.

- *Batch management server.* This provides batch management services in the DCS.

- *CM.* A CM is a regulating device or state-driven device or combination of both that is contained in a single module and can carry out basic control.

- *EM.* An EM is a functional group of equipment that can carry out a finite number of specific minor processing activities.

- *Arbitration.* Unit-owned CMs and EMs should have a fixed assignment to a USCM.

Implementation of the USCM is depicted in Figure 18.5.

Implementation Illustration

The Unit Control environment of the ISA-88.01 model is implemented in the DCS controller or application control environment of the DCS. Implementing the USCM with the procedure executing in the controller offers the following advantages:

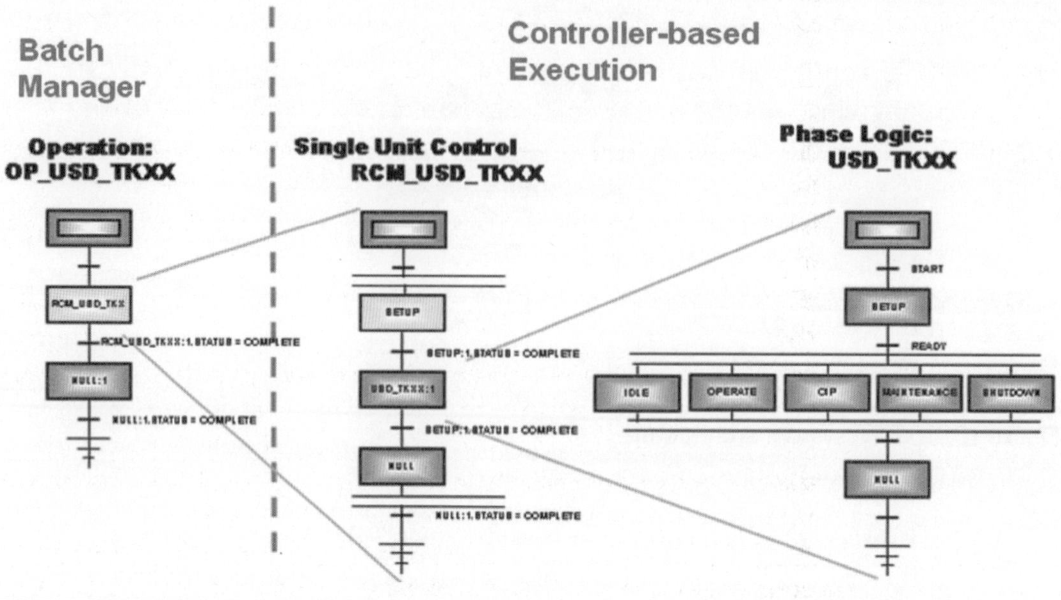

Figure 18.5.   Unit shutdown module implementation using unit phase and batch server.

- The ability to run the unit recipe or operation over the redundant controller platform with the batch server or as a stand-alone

- A higher level of reliability when the manual unit phase runs in the DCS controller as compared to recipe execution in the dedicated batch server

- Somewhat faster batch execution and reduced cycle time when compared to centralized recipe execution in the batch server

- Captured events at the DCS level that can be retrieved as needed

High Hardware Availability and Common Mode Failure

Common mode failure results from a single fault or set of faults. Redundancy is the preferred option for a system to sustain common mode failure. Upon Auto or Manual switchover to the standby batch server, recipes continue to run after restarting. In the event of a single batch management server system or absence of it, batch server failure is handled using the USCM running as a manual unit phase in the DCS controller. As illustrated in Figure 18.5, the manual unit phase runs as a procedural operation in the DCS controller and handles the process conditions through exceptions and operator-initiated shutdowns. The unit phase also handles arbitration. Thus secured shutdown in the event of loss of the batch server is possible.

Batch Journal and Report Compliance

The USCM is configured to have formula and report parameters. A Procedure Analysis tool collects data from the USCM running as a manual unit phase. Thus regulatory compliance for the batch journal data is met in the event of batch server failure as well.

Closing Remarks

The traditional way of handling unplanned events is through unit shutdown handling by Hold and Abort sequences of unit phases. Although this approach meets batch control requirements, the effort results in redundant programming and testing. It also consumes extra controller memory without providing any value-added features. In the unit shutdown strategy for exception handling in batch control applications, the EM, USCM, and unit phase provide optimized exception handling over the conventional exception handling in the hold and abort handler of the unit phase. Distribution of decision logic to handle exceptions in the EM, USCM, and unit phase provides control availability, reliability, and programming flexibility and also meets batch control requirements.

A Case Study for Batch Integration in a Specialty Chemical Facility

Presented at the WBF
North American Conference,
Philadelphia, PA,
March 24–26, 2008, by

Andy Robinson
Project Engineer
arobinson@avidsolutionsinc.com
Avid Solutions,
102 Towerview Court,
Cary, NC 27523, USA

Shawn Hull
Project Engineer
shull@avidsolutionsinc.com
Avid Solutions,
102 Towerview Court,
Cary, NC 27523, USA

Abstract

When multiple plant areas, units, and products are all controlled by a single control system in a batch environment, proper control system planning, design, and implementation are crucial to the success of a project. This chapter examines a recent project completed at a specialty chemicals plant that involved upgrading and consolidating several plant areas—each making a unique product in its own island of automation—into a single control system. This presentation will explore all stages of the project, emphasizing the use of the following:

- The ISA-88.01 standard during system design
- Object-based tools to expedite development and ensure consistency during development
- Medium-fidelity simulation tools to ensure software and batch functionality before it reaches the plant floor and aid in operator training
- A single batch manager in a multiproduct facility

Additionally, the presentation will highlight successes and lessons learned throughout the project.

Overview

Avid Solutions recently completed a large control system integration project at a multiproduct specialty chemical facility in North America. The 2400 Input and Output (I/O) project consolidated six distinct processes formerly controlled by antiquated stand-alone Programmable Logic Controller (PLC) systems or local panel boards into a modern integrated control platform. The primary goals of the project included reducing batch cycle times, enabling automation of new processing equipment, and providing a single modern control system platform across the entire plant. Such a platform was expected to reduce control system engineering and maintenance costs and reduce plant operational costs, since operators could easily cross-train across the entire facility.

As is typical for many older plants, this plant had not established a solid standard for control system platforms. Thus different platforms had been installed in various islands of automation across the facility. Two of the plant areas within the scope lacked any automation and were controlled solely by hardwired panel-board style controls and single loop controllers. The remaining plant areas were controlled by older PLCs with several different Human Machine Interface (HMI) packages scattered throughout the plant areas. Each of these distinct systems required distinct operators and system-specific training. Thus cross-training was cost prohibitive, and the ability to monitor and control the plant from a single location was not possible.

Design Approach

In conjunction with the customer, a design approach was developed to integrate the islands of automation into one system by using modern standards for batch system architecture, design, and implementation and by applying the latest technology to field devices and bus systems.

ISA-88.01 Approach

As part of the project's conceptual design phase, the customer had adopted the ISA-88.01 framework to provide a consistent methodology for batch system design and implementation (Fig. 19.1). The ISA-88.01 physical model was used to organize all aspects of the project including process cells, units, Equipment Modules (EMs), and Control Modules (CMs). The ISA-88.01 enterprise physical model was not utilized, since the plant's product lines do not extend beyond the local facility. Since the physical model was separated from the procedural model, hardware and device-connectivity design could begin without intimate knowledge of the recipes or procedures that each piece of equipment would be performing.

Procedural development was accomplished by defining activities that EMs, operations, and recipes would perform and documenting these activities in a detailed design description that would later be used to implement the software

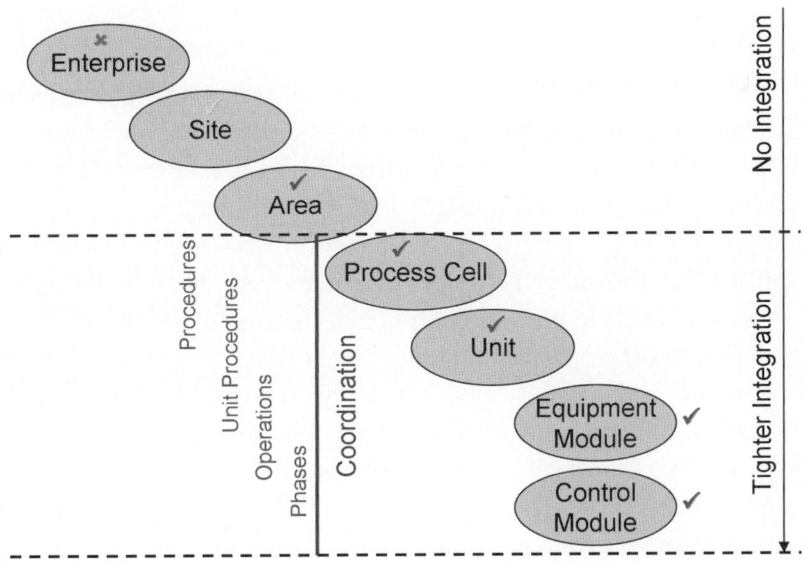

Figure 19.1. ISA-88 structure.

code. In most cases these activities or procedures were extracted from existing control code or from documented sequences of operation and then refined to ensure that they fit into the new ISA-88.01 architecture. To facilitate reviews, detailed design specifications were written in "pseudocode," allowing all audiences to review the documents without specific knowledge of the control system code. The use of pseudocode also made the documents relevant for any brand of control system and provided reference documentation for the facility personnel once the system was installed.

Procedure Development

Recipe procedures were based on activities carried out at the process-cell level, which typically consisted of multiple stages and actions that represented the steps needed to produce a single product in addition to any coproducts. Operation procedures were developed based on process units that generally consisted of major pieces of process equipment such as reactor vessels, crystallizers, or centrifuges. Operations consisted of multiple phases. In addition to phases, additional equipment phases required by EMs were developed. The general design philosophy during procedure development used an "operation-centric" approach that assigned development teams to individual process areas. This allowed multiple design teams to work in parallel, thus reducing overall development time and allowing teams to proceed without the need for a completely designed system.

Phase Development

During phase development, phase design was kept as simple as possible so that phases could be reused, would be easy to troubleshoot, and would allow for robust hold recovery without complex error routines. Previous batch projects in the plant had implemented larger, more complex phases that had proven troublesome and difficult to maintain. As a general rule, phases were defined based on a single verb associated with the activity to be performed such as "charge," "heat," or "transfer." Also, each phase was developed to be able to run stand-alone. In most scenarios, phases would be run as part of a larger batch operation that coordinated the phases; however it is sometimes necessary to run a phase manually outside of the operation. With this idea in mind, each phase was designed to handle its own equipment preparation or phase initialization logic, as opposed to assuming that equipment would be in the proper state based on prior phase execution. The phase design approach simplified phase logic implementation and also provided the ability to use the smaller, simpler phases to build flexible operations and in turn flexible yet robust recipes.

EM Development

Another area where ISA-88.01 principles were applied was during the definition of EMs. In conjunction with the customer, each EM was defined as a group of equipment that carries out a specific set of tasks independently that may be shared by multiple units. In cases where EMs could service multiple units, arbitration was used to determine ownership of the EMs, thereby requiring a unit to "acquire" the EM before it could issue a command. Typical EMs included tank and pump systems, jacket heating and cooling, and process headers serving multiple units. EMs were also designed so that they could be directed either by phases or by operator action but not at the same time. This design allowed the operator to perform complex tasks on auxiliary systems with a single command to the EM while outside the batch environment.

Modern Bus Systems and Specialized Interface Modules

When new automation was added to the existing system as part of the project, modern bus systems and devices were selected for installation. Foundation Fieldbus was the protocol of choice for new analog instrumentation. ASi bus was implemented for control of discrete valves and new process sensing switches. Profibus was implemented for interfacing with motors and variable frequency drives. The use of the newer bus systems decreased installation cost and provided the operating areas with additional process and diagnostic information that was not previously available.

In cases where the existing I/O systems were still reliable and supported, interface devices were used to connect these legacy systems to the new system. This approach allowed the customer to retain their capital investment in the original I/O system while migrating control features to the more modern and more reliable process controllers associated with the new system. The use of I/O interface devices also removed the need for rewiring and termination of hundreds of field devices, which would have required additional testing had the interface modules not been implemented. This decision avoided additional capital costs, greatly reduced startup check-out costs, and shortened the entire project execution time.

Implementation

Phased Implementation Strategy

Due to the large size of the project, we used a proven phased implementation strategy to successfully complete the implementation stage of the project. The team first categorized all the existing instrumentation and field devices to provide a logical grouping of similar devices. CMs—the base element in the control system—were created for each major category, such as valves, motors, analog controllers, and so on. The CMs were developed to be robust enough to handle minor differences between types within a given category, thus resulting in a limited number of CM types to maintain and program. The same approach was used to develop EMs and then phase modules, resulting in a full software library of tested objects that was in place before any logical and batch programming occurred. For large automation projects, this building-block approach is crucial to efficient and consistent implementation.

Object-based Development

The use of object-based development tools played a key part in maintaining efficiencies on this large project. Object-based tools allow a master object or template to be created, and then instances can be created from the master object (Fig. 19.2). It also allows the programmer to make changes to a large number of objects once they are instanced by simply changing the master object. Object-based development was used to develop CM templates, EM templates, phase templates, and unit templates based on the software building blocks discussed in the previous section. As a result, software development time was reduced since a master object could be instanced easily by changing only a select number of parameters on the instanced object. Object-based development also resulted in a more consistent product since multiple developers pulled from one master object library during software development.

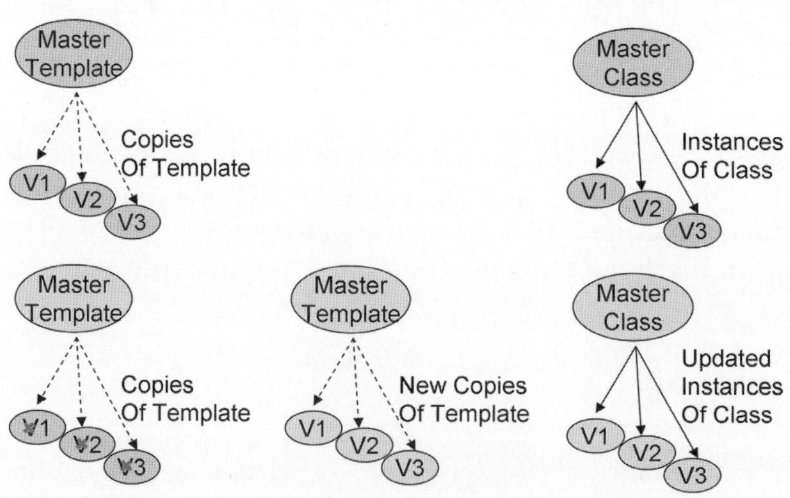

Figure 19.2.   Object templates.

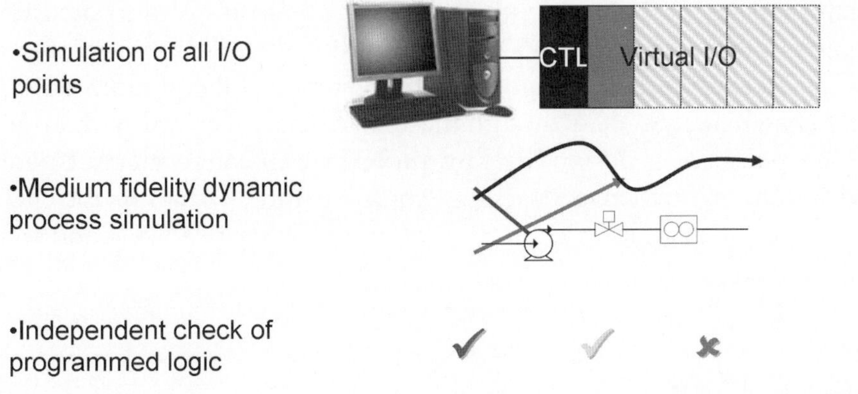

Figure 19.3.   Simulation tools.

Simulation Tools

A process simulator package was used for training and employed to allow for simulation of both I/O (traditional tie-back simulation) and the physical processes. The simulation package provided the ability to fully configure the offline system in the same manner as it would be installed in the field. This approach allowed for nonintrusive, fully functional configuration, testing, and debugging of the production software prior to field installation.

The simulator package was also used as a training environment for control system operators and engineers. By having a simulated process, the operators could execute entire batches just as they would on the live system, once installed. Using a simulation package not only dramatically reduced errors found during startup but also permitted the plant operators to comfortably take control of the system once it was turned over to them, due the extensive real-life training acquired during the project phase.

Single Batch Manager

One major challenge in this project involved the use of a single batch manager for the entire facility, which already had other units actively running. One major drawback of the system was that adding a new phase to a recipe or a new parameter to a phase required the batch manager to be stopped and restarted, which impacted the entire plant, as all running procedures had to be put into a hold state. This scenario illustrates the importance of properly designed phases and robust hold recovery logic to enable phases and batches to be held and restarted without losing a batch or causing severe operational problems. Another problem with having a single instance of a batch manager is that one problem on one computer could potentially impact every batch in the system and possibly bring the whole facility to a state where no batches can be executed. Unfortunately the system architecture did not provide for redundant batch managers. Although these problems do exist, the customer felt that through proper system design, backup procedures, change management scheduling, and system maintenance, the issues could be mitigated. Thus the client felt that the advantages of a single batch manager (including reduced cost and reduced system maintenance) outweighed the disadvantages.

Project Successes

Several successes were realized during this project, resulting from the design approach and implementation techniques highlighted in this chapter. Most notably, all the primary goals of the project were reached. All the plant areas were started up safely and per the project schedule (with only minor errors), which allowed the customer to start making usable product after only a few dry runs of the new system. In addition to the successful startups, in almost all cases batch cycle times were decreased due to the increased automation. A single operator could now comfortably monitor multiple processes from a single location, which allowed the plant to free up multiple operators to perform other essential duties. Additionally, due to a well-conceived implementation structure, the system was delivered with a configuration that was easily adaptable to future recipe or process changes, making it easier to maintain and implement changes.

Applying ISA-88.01 to Software Migration Engineering

Presented at the
WBF North American
Conference, Atlanta, GA,
March 5–8, 2006, by

Allen D. Benton
Engineering Partner
adbenton@prime-integration.com
Prime Integration,
3217 Brampton Street,
Dublin, OH 43017, USA

Abstract

Dow Chemical maintains thousands of batch control systems worldwide built on its proprietary MOD 5 computer system. Over the past decades, they have undertaken meticulous efforts to preserve and enhance these systems. Dow intends to preserve and reap the rewards of this diligence by developing best practices and tools for reverse engineering to migrate this intellectual property to next-generation platforms. In the summer of 2005, a proof-of-concept project began with the following goals:

- Improve design recovery to ensure functionality is preserved with a high degree of confidence

- Improve design documentation to accelerate the migration process from concept to implementation

- Import software into relational and hierarchical (XML) databases

- Analyze and map imported data to an ISA-88.01-based framework

- Refine data artifacts based on reuse potential and export to Detail Design Specifications (DDS)

- Develop functional narratives based on software patterns and export to Functional Design Specifications (FDS)

This chapter will (1) provide an interim report on the results of these efforts, (2) introduce a common terminology for understanding software migration engineering, and (3) show how these steps can be applied to other batch control systems.

Introduction

In the early days of process automation, Dow engineers, mathematicians, and scientists programmed small computers to solve complex control problems that they understood intimately. The resulting "super code" is Spartan, yet rich with process and safety engineering knowledge. As pioneers in the field, Dow developed and maintained their own proprietary systems and software. Thirty years later, they are faced with the necessity to mine this software on their own because most commercial Process Automation System (PAS) vendors, with their object-based application frameworks (i.e., "super-sized code"), are wholly focused on the needs of their installed base. The research looking to span this architectural divide with software translation is still basic and unproven.[1] The good news is that research directed at design recovery has demonstrated tangible ways to supplement what has been a labor-intensive and often fruitless activity. The starting point for Dow and others is not to translate old software into new but to reinterpret it to derive the original intent that we know as "requirements." Dow is committed to advancing this new engineering discipline by developing competent, committed employees; flexible, proven processes; and light-weight tools to reach their objectives.

System Background

In the early 1970s, Dow developed a fairly rapid succession of MOD 1, MOD 2, and MOD 3 computers under the direction of computer scientist Wayne Depree. A single all-digital MOD 3 control system was installed and commissioned in Midland, Michigan, using a minicomputer with just 16 KB of memory. This system incorporated an early version of the Dowtran process control language to replace earlier assembly language programming. It was able to execute two sequences per Input/Output (I/O) subsystem (referred to as a "can") with a maximum of five cans per system. In 1974, the MOD 4 was introduced, which contained a 16 bit processor and a commercially available computer with 80 KB of memory. It interfaced to a SY FA Process Information system that provided rudimentary history and reporting capabilities. A total of seventy systems were successfully built and installed over the next four years. The first MOD 5 computer prototype was completed in 1978. At its core was a military-grade computer built by Plessey Semiconductor, originally conceived to program tank turrets to fire while moving. Depree customized the basic CPU architecture to improve processor throughput and hardware reliability. This system became the first fully redundant process control system with parallel CPUs, I/O, and power supplies. It also allowed programs to be reloaded without having to shutdown operations. In 1980, the MOD 5 went into full-scale production. In the decades since, the Digital Equipment Corporation line of minicomputers was incorporated into the system architecture, starting with the PDP-11/40, followed by the VAX, and currently running on the Compaq Alpha. A typical MOD 5 installation consists of five cans. Each can has a capacity of 400 I/O, graphical displays, and operator interface units. The current MOD 5 system is certified for SIL 3 applications by the German certifying agency, TÜV.

Software Background

An important feature of the MOD 5 control system is that every sequence is executed by the system's computers at a rate of once per second, with options to execute selected portions at rates of ten to one-hundred times per second. A key characteristic of Dowtran programming is that each statement of a sequence is independent (no nesting) and sequential (no branching) and is conditionally executed each time the program cycles. This creates a true deterministic programming environment similar to Programmable Logic Controller (PLC) logic.

The Dowtran language and database use an array of elements for all real-time data (I/O and non-I/O). There are about twenty regulatory element types that support instrument I/O, alarms, display inputs and outputs, redundancy, and so on. The Dowtran language and database also include support for batch applications in the form of generic recipe element and step event arrays. All arrays are global in scope, which complicates translation to the ISA-88.01 model because of the opportune use of cross-unit coupling.

Migration Terminology

To begin to understand migration engineering, it helps to start with a model that most everyone can agree on. The Software Engineering Institute (SEI) has a model they call the "horseshoe"[2] that attempts to divide the reengineering process into phases (recovery, transformation, and development). The horseshoe name is derived from the arc shape that appears on a graph of abstraction level over the life cycle of a project. One difficulty of this model is that the user must first adopt an architecture-centered approach to software design, which is still in research. An important aspect of the model is that it recognizes the pragmatic reuse of software artifacts across the reverse–forward engineering divide.

Figure 20.1 presents a new model that builds on this work but focuses on the deliverables associated with classic software design. The model recognizes the adage that "reverse engineering is forward engineering in reverse."[3]

Reverse Engineering

The first leg of the process (A→C) is referred to as Reverse Engineering (REV-ENG). Starting with computer-readable configuration files, engineers advance from point A to point B using tools to create human-readable DDS containing data elements, data flows, and control flows. These documents are patterned after the modularity and data structure of the existing system (I/O, graphics, sequences, recipes, etc.). During this step, meaningful descriptions are applied to the data, including engineering units of measure. Once these documents have been reviewed and approved, the next phase (point B to C) starts with creation of an ISA-88.01 representation of

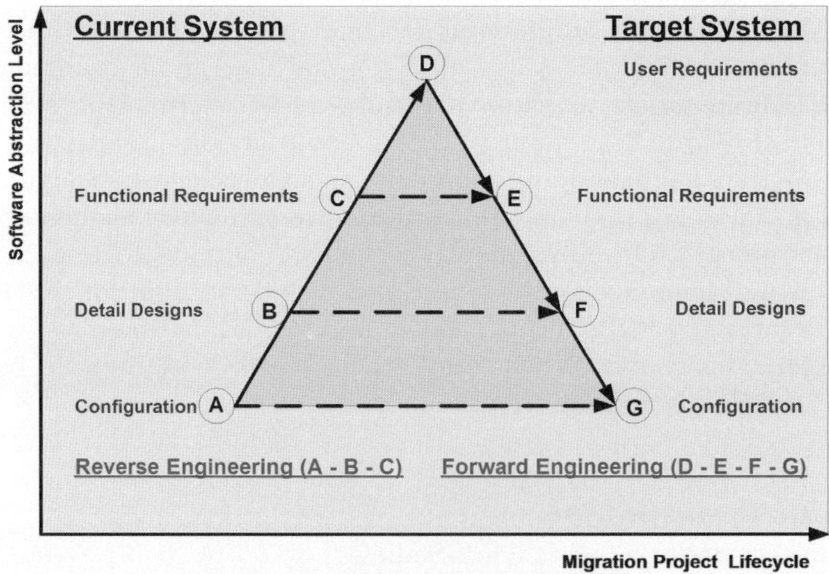

Figure 20.1.   Migration engineering model.

the process domain that the software currently operates in. Many assume this is not possible if the system was implemented without the aid of these models, but in reality, the absence of modularity allows the existing configuration to be manipulated quite easily. This phase is completed with the creation of the FDS, which is also reviewed and approved. This document is made up of brief narratives describing the role and logic of each useful software artifact at the Unit, Equipment Module (EM), and Control Module (CM) levels of the Physical model, as well as at the Unit Procedure, Operation, and Phase levels of the Procedural Control model.

Forward Engineering

The second leg of the process (D→G) is referred to as Forward Engineering (FWD-ENG) to distinguish it from traditional software engineering. This is a useful distinction, since classic software engineering tends to be iterative and labor intensive, while FWD-ENG emphasizes document automation and software reuse. The converging arrows ([D→E, C→E], [E→F, B→F], and [F→G, A→G]) show where reuse typically occurs. Data from REV-ENG documents is combined with abstractions from the FWD-ENG deliverable to create a useful starting point for the next deliverable. Tools and a data repository can be developed to automate each phase of this process. FWD-ENG starts with the user spelling out, in general terms, what existing functions are to be removed and what functions are to be added or changed. This document is referred to as the User Requirements Specification (URS) and is also reviewed and approved. The remaining documents and the processes used to create them are mostly self-explanatory.

Why bother with multiple steps? In a word, the answer is "semantics." Each step of the process introduces generalization or abstraction into the system description. Abstraction is what allows the essence of the current system to be captured and carried over to the target system. The process is optimized when software artifacts are abstracted only to a level where the semantics that describe them apply also to the target system. Semantic distance describes the degree to which an abstraction is different from its implementation.

Static Analysis Techniques

With full access to a running system, design recovery could be accomplished by observing all behaviors under all possible conditions (Dynamic Analysis). The reality is that most systems are continuously online, and many scenarios have safety implications. The techniques known as Static Analysis rely on source code and configuration data to answer the questions, "What is the system capable of doing?" and "How does it accomplish it?" In the same way Google was able to capitalize on the insight that Web pages contain valuable structural and linkage information, static analysis has demonstrated that source code is equally rich in this type of content and can be leveraged to make reverse engineering projects more cost effective. The proof-of-concept work started with the gathering of industry-proven approaches to the understanding of a software system from a functional perspective.[4] The existing MOD 5 software migration toolkit supports many of these analyses. The following list describes some currently state-of-the-art techniques:

- *Configuration Disintegration.* This technique starts with a language-specific parser to break source code into tokens that are saved in a relational database as data elements, data flows, and control flows. Certain data elements of interest (e.g., Analog Outputs, Digital Outputs, and Step transitions) are then recombined into "Clusters"[5] using Dependency graph techniques that trace flow relationships between program statements. This step effectively removes language syntax variations so general purpose tools and database structures can be shared for further analyses.

- *Pattern Matching.* This technique compares clusters using distance algorithms to look for like repetitions of data flows and control flows.[6] A comparison algorithm that exploits knowledge of the software grammar is able to point out differences quite accurately. Patterns are useful for recovering design solutions within a process domain that only need to be described, reimplemented, and tested once at the functional level.

- *Program Slicing.* This technique allows the user to view clusters in the context of the original source code.[7] An XML representation of the source code is generated and XSLT is used to create an HTML-based source code listing. Hyperlinks are inserted in the source code to allow the engineer to traverse the tree structure of the cluster.

- *Dependency evaluation.* This technique takes a macro view of the system by showing the interdependence of data elements with different subsystems such as operator displays, recipes, I/O, reports, and so on. Traditional implementations of this technique include cross-reference reporting.

- *Metrics Evaluation.* Once clusters have been assigned to CMs (Physical Model) and Phases (Procedural Control Model), various characteristics such as size, complexity, quality, maintainability, and so on can be measured[8] (e.g., the Halstead Complexity factor and

Design Structure Quality Index) and assigned at the cluster and module level. These techniques can be used to assess reimplementation and testing risks or to highlight overly complex control strategies.

- *View Exploration.* No single view of software is best when attempting to analyze functionality. This technique makes at least four views available: (1) Physical Hierarchy view, (2) Procedural Hierarchy view, (3) Statement Type Cluster view, and (4) Source Code view. In addition, each view should provide easy navigation within the view and to other views. Each of these views also needs to be available in paper report formats that can be printed and marked up during review meetings.

- *Flow Visualization.* This technique differs from Exploration in that it uses a graphical representation of data flows (data flow diagrams) and control flows (state transition diagrams) to represent portions of the system. A useful chart type for batch applications is the Step-Transition chart, which shows each step and all possible transitions to and from the step.

Migration Process

Figure 20.2 shows the stages of analysis, the techniques applied, and the deliverable products. Four sequences selected from different processes were used to exercise the proof of concept. The following steps have been formed to provide the most efficient path through the REV-ENG portion of the migration process:

Stage 1.   Design the audit.

Step 1.   Take inventory of existing design documents.

Existing design documentation may be terribly out of date, but it often offers insights into the developer's intents. The revision history of these documents (if available) often suggests opportunities for design improvement. It can also reveal the implicit modularity of the process domain that monolithic software was written to control. Only certain documents (e.g., piping and instrumentation diagrams and instrumentation lists) need to be updated and verified to become useful reference documentation for the physical instrumentation hardware.

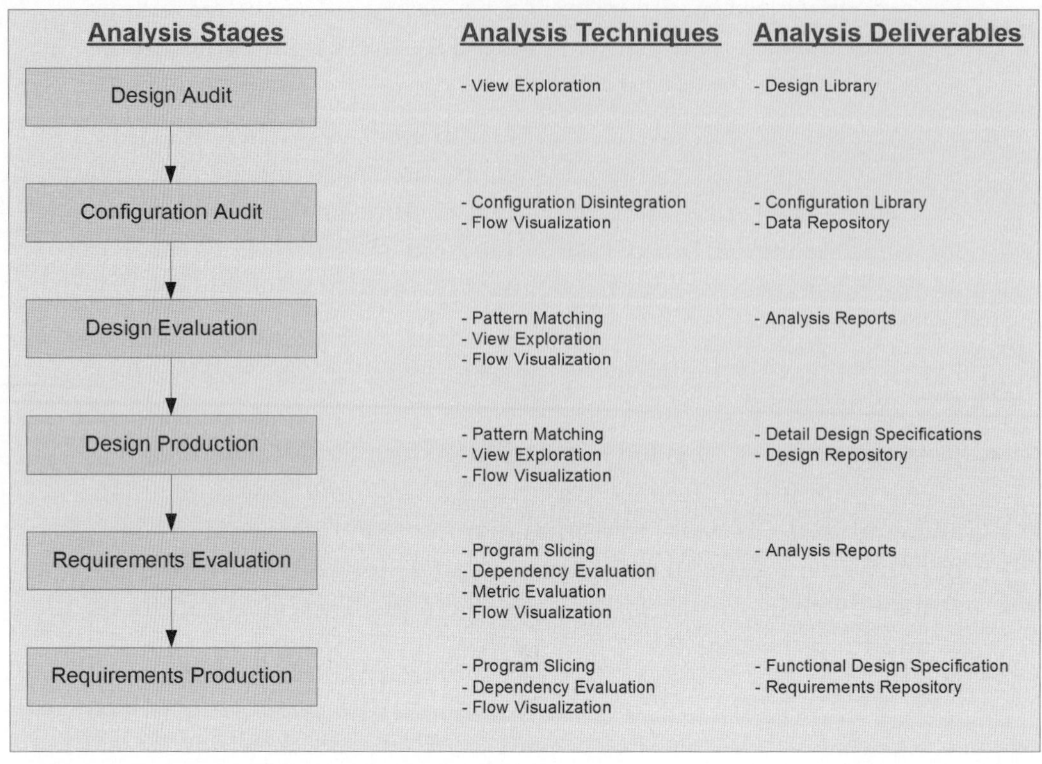

Figure 20.2.   Reverse engineering analysis stages.

Stage 2.   Audit the configuration.

Step 2.   Take inventory of existing software configuration.

All application software and data need to be identified and cataloged with names that describe the functionality within. These files should be in ASCII format and accessible by the computer where the analysis work is being done. If ongoing software maintenance is taking place on the original system during the migration project, detailed change control procedures should be in place to ensure changes appear on the target system as well.

Step 3.   Import software into a data repository.

Depending on the structure of the source files, various techniques can be used to parse code into tokens and write it to various tables in a relational database. Ideally, this database would be accessible from a central server that supports concurrent users. The only requirement for the source files is that the grammar be context-free and that the source code must compile without errors. In Dow's older installations, source code comments are the richest source of design documentation so one of the requirements for tool development was to preserve comments and make them accessible in the correct context.

Stage 3.   Design evaluation stage.

Step 4.   Modify (complete or correct) data definitions.

The importance of this step is tempting to overlook since the existing system works. The goal is to clean up and standardize all data element descriptions, instrument ranges, and instrument units of measure to create a "data dictionary" that can be verified and referenced later. Equations, scaling, and constants need to be understood and checked for invalid or outdated assumptions. Alarm types and limits should be reassessed for present-day relevance. If this stage is bypassed, the result will be unnecessary testing, perpetuated quality problems, and nuisance alarms.

Stage 4.   Design production stage.

Step 5.   Categorize data by reuse potential and ISA-88.01 hierarchy.

Ideally, assessing reuse potential is a task reserved for the FWD-ENG leg and creation of the URS. In practice, decisions need to be made at the design level as to which software artifacts to further analyze at the functional level. The decision to assign or not assign clusters and parameter data to the ISA-88.01 hierarchy effectively accomplishes this task, since only linked elements are visible at the functional level. The result is an extension to the ISA-88.01 hierarchy, as shown in Figure 20.3. This drawing is not intended to suggest an extension to the ISA-88.01 specification but simply to provide a frame of reference.

Step 6.   Export to DDS.

Dow recognized that obtaining buy-in from future stakeholders for the new system needs to begin early. One way to accomplish this is to bring them into the review and approval process of the

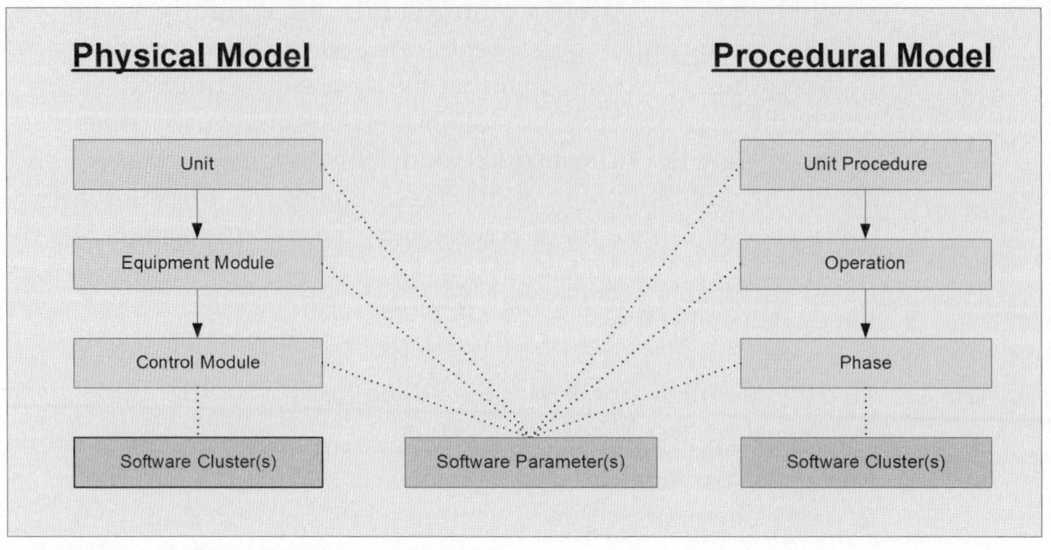

Figure 20.3.   CM and phase linkages.

design documents. In this context, a wealth of information about how the existing system operates in the real world will begin to surface. This feedback must be captured in an issues database and analyzed in order to establish credibility with the team. One obstacle that was addressed was the need to present detailed documents in the format Dow engineers had grown accustomed to through their existing work processes. A Microsoft Visual Basic for Applications (VBA) conversion utility was developed to translate the database-optimized detail design document to a reviewer-optimized version.

Stage 5.   Evaluate the requirements.

Step 7.   Develop upper-level narratives (units, EMs, unit procedures, and operations).

This is a top-down approach to requirements development that defines the scaffolding (context) on which the detailed system requirements will be attached. The task is especially important if the existing system was created without an ISA-88.01

perspective. Numerous educational resources are available from WBF and ISA to guide this effort.

Stage 6. Produce the requirements.

Step 8. Develop functional narratives for CMs and phases.

This is a bottom-up approach to requirements development. The task is to synthesize functional narratives from cluster-level source code. A useful narrative melds a process description (purpose) with a software description (automation scope) that communicates the original intent to multiple reviewers. For complex requirements with nested decision branches, a pseudocode vocabulary needs to be established and enforced until it becomes natural.

Step 9. Export to FDS.

This is the second opportunity to promote stakeholder buy-in. If requirement narratives have been captured in a relational database, tools can be used to convert the document into a Microsoft Word format using a customer-defined template. As with the previous review, new information about the existing system will surface, and this feedback must also be captured and analyzed.

Migration Tools

Figure 20.4 shows the organization of views used to implement a proof-of-concept Software Migration Environment (SME). Microsoft Access was used as the database and application environment for speed of deployment. A single Dowtran sequence may vary in size from 1 to 6 KLOC (excluding comments), which can result in a database in the order of 20 MB. These file sizes create a practical limit of one unit design per database for the most complex applications. Potential workarounds include using a more efficient data schema or upgrading the database engine to a Microsoft SQL Server Desktop Engine (MSDE). Also, database partitioning is not considered a handicap because the normal division of engineering labor occurs along unit boundaries. Database linking could later be used in the future to generate cross-unit reports for project-wide analysis.

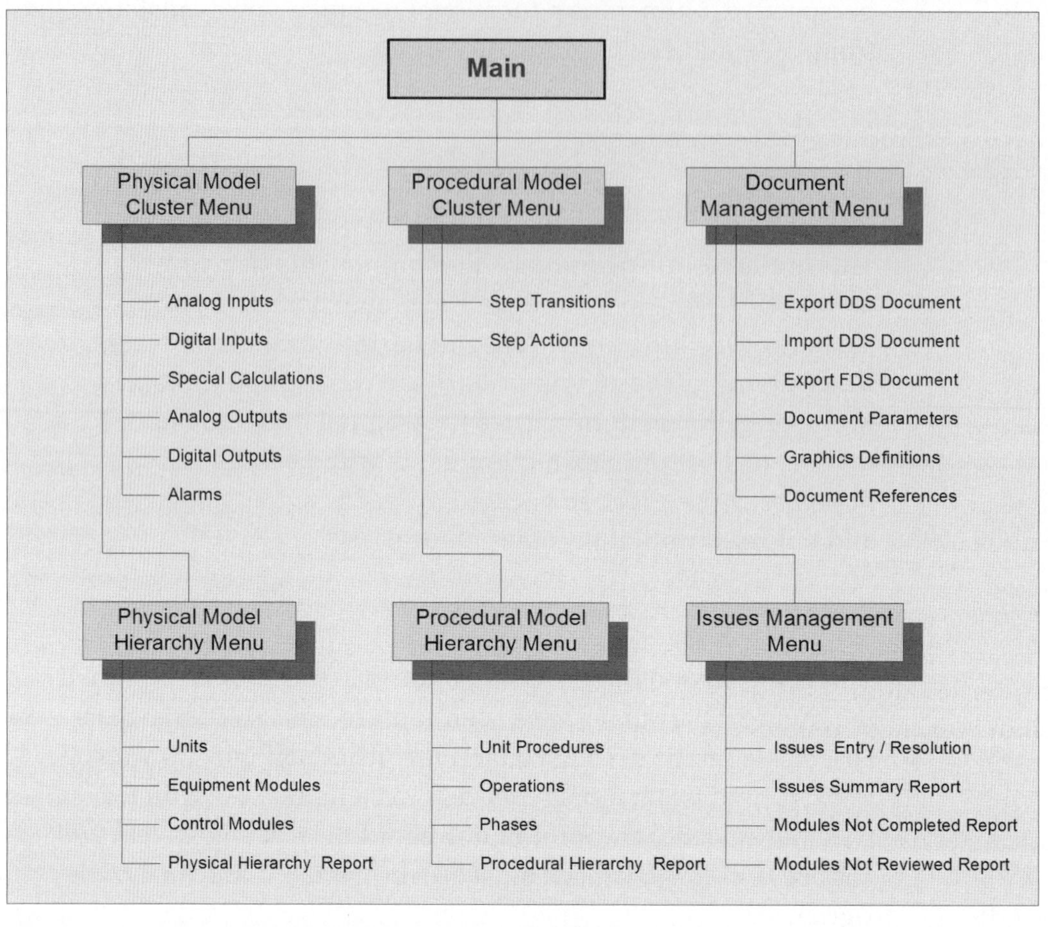

Figure 20.4.   Software Migration Environment (SME).

Summary

Dow has taken a robust first step toward mining its MOD 5 applications using these proof-of-concept reverse engineering tools and processes. They have demonstrated that design recovery can be systematically achieved using industry-proven software migration techniques. By applying the ISA-88.01 hierarchy, they have imposed modularity and demonstrated conformity to common object-oriented batch frameworks with the intent to accelerate software design and development. An equally important next step will be to approach commercial PAS vendors with electronic design documents and design data repositories to determine which forward engineering tools will be made available to automate significant portions of the configuration process.

References

1. Harsu, Maarit. 2000. Identifying object-oriented features from procedural software. Department of Computer and Information Sciences, University of Tampere, Finland.

2. Carnegie Mellon University Software Engineering Institute. Reengineering: The horseshoe model. http://www.sei.cmu.edu/reengineering/horseshoe_model.html (accessed 2005; site now discontinued).

3. Baxter, Ira D., and Michael Mehlich. 1997. Reverse engineering is reverse forward engineering. Proceedings of the Fourth Working Conference on Reverse Engineering.

4. Jin, Dean, and James Cordy. 2004. *A service sharing approach to integrating program comprehension tools*. Kingston, Canada: Queen's University.

5. Anquetil, N., and T. C. Lethbridge. 1999. Experiments with hierarchical clustering algorithms as software remodularization methods. Proceedings of the Sixth Working Conference on Reverse Engineering, October 1999.

6. Baxter, Ira D., Andrew Yahin, Leonardo Moura, Marcelo Sant'Anna, and Lorraine Bier. 1998. Clone detection using abstract syntax trees. Proceedings of the International Conference on Software Maintenance.

7. GrammaTech Inc. 2000. CodeSurfer technology overview: Dependence graphs and program slicing. White Paper, GrammaTech Inc.

8. Seviora, Rudolph E. Lecture notes related to metrics (source code). http://www.swen.uwaterloo.ca/~kostas/ECE750-3/metrics-1-notes.pdf.

Automating the Manufacture of Highly Energetic Organics Using the ISA-88.01 Models

Presented at the WBF
North American Conference,
Woodcliff Lake, NJ,
April 13–16, 2003, by

John Arnold
Director of Automation
arnoldja@tricon.net
Cochran Corporation,
2205 Nantucket Street,
Johnson City, TN 37604

Dennis Brandl
Chief Consultant
dnbrandl@bellsouth.net
BR&L Consulting Inc.,
208 Townsend Court, Suite 220,
Cary, NC 27511

Abstract

The manufacture of Highly Energetic Organics (HEOs) consists of three distinct and physically separated process stages: Reaction, Wash, and Crystallization. Due to the reactive nature of the products produced, each step is performed in separate and isolated manufacturing areas. In 2000, a project was initiated to automate the complete production of the manufacturing process and also to install state-of-the-art automation in all three of the manufacturing steps. The effort was completed in late 2001 and has been in use for almost six months at the time of this writing. One of the most visible signs of success of the new system has been the reduction of off-specification products from 15 to 20% to less than 2% of the batches produced. This chapter addresses some of the lessons learned and describes some of the trials, tribulations, and issues involved with bringing batch automation to a process that has traditionally been handled manually. Some of the issues revolve around process measurements, control problems, the replacement of human actions with computer actions, and resistance to change. This chapter will discuss how ISA-88.01 concepts were applied in the manufacture of HEOs.

Introduction

In the world of process control systems, one of the most dangerous operations is the manufacturing of explosives. This chapter deals with some of the issues that can arise while manufacturing HEO materials using the ISA-88.01 models. Production of these materials is inherently a batch process and consists of three distinct and physically separated process stages. These stages are Reaction, Wash, and Crystallization. Due to the reactive nature of the products produced at each stage, each stage is usually performed in separate and physically isolated manufacturing areas in a plant. In 2000, a project was launched to automate the complete production of the HEO manufacturing process at an existing facility. The project also involved the installation of state-of-the-art control systems onto all three of the manufacturing steps. The effort was completed in late 2001 and reduced the production of off-specification products from 15 to 20% to less than 2%.

There were multiple lessons learned during this project, but three of them stand out as issues that can arise in many batch automation implementations. These issues were (1) poor management reaction to the perceived cost of batch automation, (2) unrealistic expectations of automation, and (3) unrealistic expectations of batch automation reuse. Some of the problems occurred because the project brought batch automation to a process that had traditionally been handled manually, and therefore there was no history of batch automation projects. The three main lessons also relate to the three main steps in the production process, which are as follows:

- *The first stage: Reaction—A management reaction to batch.* The lesson learned was that if you do not purchase a batch executive add-on, recipe management is still required in almost any batch scenario.

- *The second stage: Wash—Automation doesn't replace the common senses.* The lesson learned was that automation of manual operations requires you to formalize common sense rules and procedures, and replacement of simple observation with sensors is not always as easy as it looks.

- *The third stage: Crystallization—No two systems are exactly the same.* The lesson learned was that seemingly small differences from a management viewpoint can be large from the design and implementation viewpoint and need to be planned for.

Reaction: A Management Reaction to Batch

Due to the volatile nature of HEOs, each stage in the manufacturing process is performed in distinct and physically separate buildings. This separation is designed to reduce the possibility of an accident in one process area affecting production in another area. Figure 21.1 illustrates the general plant layout. The general process description is illustrated in Figure 21.2; the detailed process definition is not shown due to concerns about displaying the production methods for high explosives.

The first stage in the manufacturing process is the Reaction stage. This stage is the most sensitive due the highly exothermic nature of the reaction. The reaction is also very rapid; the first step in the process is complete in less than an hour. This stage in the process has been automated for many years because of these considerations.

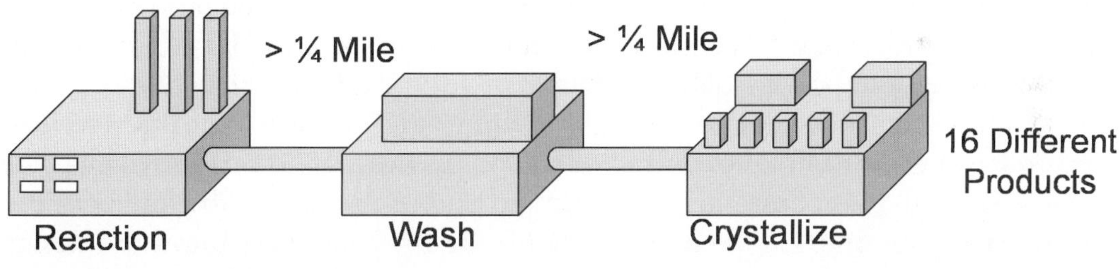

Figure 21.1.   General physical layout.

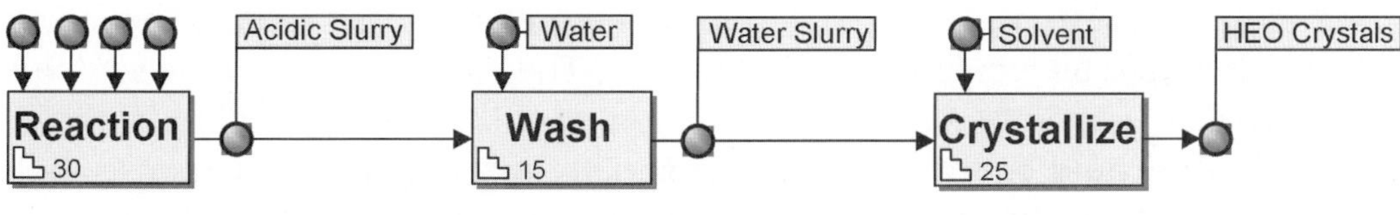

Figure 21.2.   Simplified general recipe.

The present automation system is the third one deployed in the last thirty years. The first generation of automation was installed in the 1970s and was based on single loop controller technology. The second generation was developed in the 1980s using hybrid Programmable Logic Controllers (PLCs). Unfortunately, the automation solutions developed in the 1980s were highly inflexible. While it was possible to switch from one product to another using this technology, it required the complete installation of another program in the controller. Also, as a result of the system architecture, the simultaneous manufacturing of products within the building was impossible. The 1980s solution predated the release of the ISA-88.01 batch standard.

The current generation of automation is based on a new Process Automation System (PAS). Design requirements of the system allow for both the simultaneous manufacturing of products as well the ability to rapidly set up either of the two reaction trains to manufacture any product. There are sixteen final product recipes in the system. When intermediate formulations are included, the total number of recipes in the system is twenty two. At any point in time there may be up to thirty simultaneous and different product recipes active. Since all the products had different manufacturing recipes, it was necessary to develop a batch executive that could execute within the controller. This batch solution had to be developed using the standard components of a basic PAS since a management decision was made very early in the system specification process that a batch executive add-on was not necessary and would not be purchased. One reason for excluding a batch add-on in the project was because the end user did not see a need for a sophisticated and expensive batch package. The packages on the market today tend to impose an approach on the end user. Cost, complexity, rigidity, and support issues lead to the end user's decision to build a batch solution using standard components in the off-the-shelf PAS. This decision moved some of the expense from capital investment to project manpower. Without the batch add-on, the construction of recipe control, phase control faceplates, command forms, and a phase logic interface had to be undertaken upfront. The contractor essentially developed a data-driven batch package because the requirements demanded one, even though the end user did not see a need for one. However, using the ISA-88.01 models reduced the design time and allowed the system to meet its flexibility requirements.

The batch package was developed to execute in the controller using standard components of the PAS. The recipe system consisted of data tables ("recipes") and multiple execution engines that interpreted the data in the tables. The recipes and batch engines all resided in the controllers. The tables contained the setpoints for the phases, setpoint selection information, and the execution pattern for the phases. Phases could be set up to execute serially or in parallel. Any phase could also be executed multiple times with different setpoints. The execution engine managed the initiation and resetting of the phases, but phase logic determined how the phase setpoints were loaded from the recipe. The recipe manager simply told the phase which set of phase setpoints were to be loaded. The operator could specify the setpoint values, or the recipe could contain the setpoint values.

Recipes resided in the controllers and were moved from the product recipe matrix into a working recipe table at the start of each batch. The working recipe was then used by execution engine components in the controller to process the data contained within the recipe to determine which phases to execute, which commands to send to the phases (e.g., Start, Reset, Hold, or Abort), which setpoints to utilize, and when to initiate the various phase commands.

The execution engine was composed of two graphical components—Sequential Function Charts (SFCs) and a function block that acted as a stack server. The phase list was held in a stack in the recipe data table. (A stack is simply a table of data that can be processed sequentially.) The first SFC was designated as the batch manager and was designed as a graphical "For/Next" loop that worked with the stack server to sequentially process the phase data contained within the stack of the recipe. The recipe contained both the data and the list of phase IDs (stacked) used to sequence phase execution, as shown in Figure 21.3.

The second SFC of the execution engine was designated as the phase manager. The job of the phase manager was to determine which phase was called for by the current stack output, which phase command was to be sent to the phase (if the phase was to be started), and which set of setpoints was to be used, and finally to send the appropriate command to the phase. The phase manager also

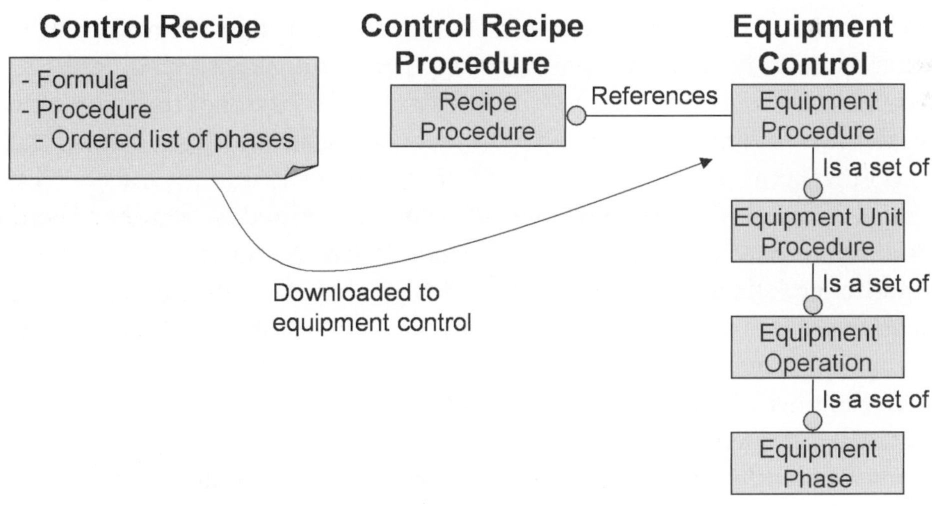

Figure 21.3.   Phase sequence execution controlled by equipment control.

was designed to wait for the appropriate response from the phase before returning control to the batch manager.

The advantages of this approach include its simplicity, flexibility, and low cost. Using this approach allows sequences to be restructured on the fly without any programming effort, meeting the primary goal of the ISA-88.01 models. This is particularly valuable during checkout, when it is often necessary to quickly restructure recipes for partial runs. Figure 21.4 illustrates the operator view of the executing recipe that was developed for this application in the context of ISA-88.01 structures.

One advantage of developing the batch system using the basic system components is that the entire batch interface is inherently contained within a single control system. This means that the execution and control of the batch can be controlled and monitored directly from any Human Machine Interface (HMI) without going to a separate batch HMI package. This reduces hardware costs and simplifies operator training. Operator training is particularly significant when automating a plant that has been operated manually for over 60 years. This is because many operators are middle aged and are often uncomfortable using computer systems. Any reduction in the number of systems that an operator had to use reduced the time required for the system to gain acceptance.

In addition to the problem of building a batch system from scratch, the Reaction stage was also challenging from a process control viewpoint. The batch steps and the sequences required were well defined and had been for many years. The control problem was getting the reaction started and keeping it going. The cooling system was oversized, and this made the implementation of a robust and stable temperature control difficult. In addition, all the products had different operating temperatures and rates of introduction for raw materials. These two factors alone made the design of the temperature control difficult. When the timeframe of the batch was factored into the equation, it led to further difficulties. At the beginning of the reaction it was important not to apply cooling too early or the reaction would not initiate. However, after the reaction reached initiation, it was critical that sufficient cooling be applied to prevent the reaction from running away. Without a flexible and adjustable control strategy, the reaction tended to cycle between never initiating to overheating. Complicating the problem was the fact that the strategy could not be realistically tested until the system was in production. Therefore the control strategy had to work on the first and on all subsequent batches. This was quite a challenge, and extensive use was made of a detailed simulation of the process that was developed. This simulation provided a realistic model of the entire process (including the heating and cooling system), which was used to validate the automation and control strategies. Thus even before the first water batch was performed, there had been over one hundred complete batches executed against this detailed simulation. This allowed startup to be focused more on verifying the

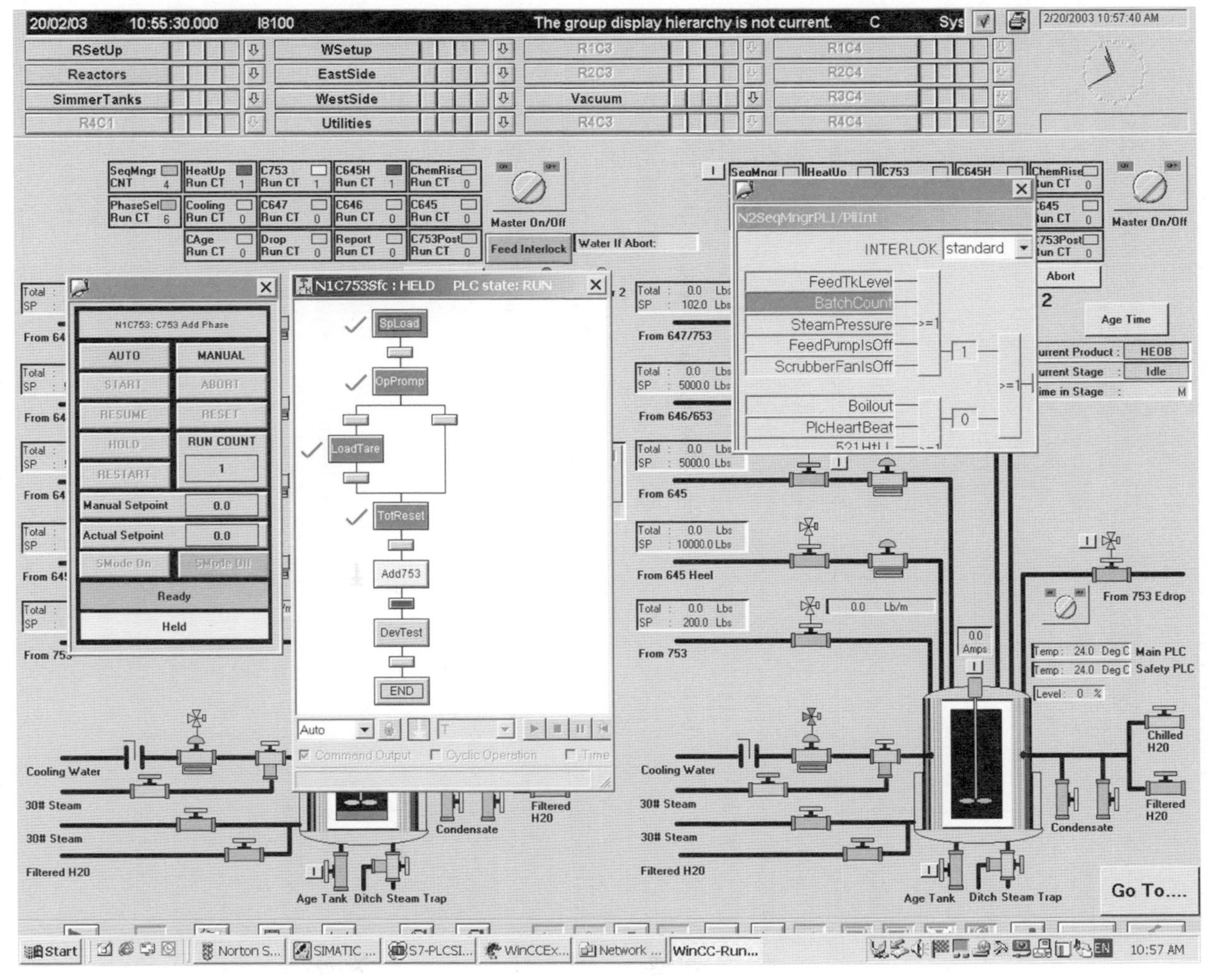

Figure 21.4.   Example of batch or phase monitoring from within an HMI package.

process and less on troubleshooting flaws in the implementation of the solution. The result was that the time required to bring the process into production was relatively short (4 weeks), considering that the process, the building, the wiring, and the instrumentation had been gutted and rebuilt from scratch.

As with the batch executive, there was no management support for investing in a simulation package. This meant that the contractor was forced to utilize standard PAS components for constructing an Input/Output (I/O)-based process simulation. This simulation included all physical devices (valves, motors, etc.) and the physical process (heat and energy balance plus reaction). Simulation was accomplished by developing a mathematical description of each and every piece of equipment within the system. The models were then programmed using the PAS programming languages and installed within the controllers along with all the automation structures. In the end, the cost of developing the simulation was minimal considering the fact that no extra hardware or software was purchased. Also, operators were not required to learn the nuances of a simulation package and its communications interface to a control system. In addition, the simulation could easily be included or excluded from the controller; thus the simulation could be resurrected at will whenever training or system expansions were needed.

Wash: Automation Cannot Replace Common Sense

The second stage in the manufacturing process for HEOs is Wash (filtration and washing). This manufacturing stage had been executed manually since the 1930s. The process had never been automated and the design requirements were not as clear-cut as the Reaction stage requirements. However, there was a significant economic benefit to be gained from automation due to the labor-intensive nature of the process. Automation could allow the building to operate remotely, thus reducing labor costs and increasing personnel safety.

When the project was started, there were significant questions concerning whether the process could even be automated. The question of building an automation solution revolved around two basic issues—pumping and sensing. Before a decision could be made as to whether or not to invest in automating this system, a major question had to be answered: Can a mechanical system be designed in conjunction with automation logic that could remove almost all the solids from the wash tank without utilizing excessive amounts of water?

Excessive water in the slurry would cause significant problems in the next step in the process. Excessive clumping on the filter cloth would age it rapidly and lead to extended cycle times. The process of getting all the material off the cloth and out of the tank while utilizing minimal water proved to be quite a challenge.

This problem was compounded by the fact that the pump that was utilized was not a self-priming pump. Keeping the pump primed proved to be quite a challenge. Many hours were spent testing and changing the pump-out phase. Eventually a successful strategy was created, but it took many test batches.

A particularly challenging task involved detecting when a "cake" formed. Once a cake is formed, it is imperative to stop pulling a vacuum on the process. Otherwise cracks may develop in the cake, and subsequent washings become ineffective due to the channeling of wash water through the cracks. This leaves too much acid in the final cake and significantly extends cycle times. Cracks are a significant problem and must be avoided if a system capable of remote operation is to be developed.

Some may suggest that cracks are relatively easy to sense due to a detectable sudden and significant change in pressure. In this process, that was not the case. Vacuum was only sensed in the vacuum header, and because there were potentially six tanks on the header, a crack in one cake was not noticeable at that location. Some might also suggest that cracks are easily corrected. This might be true for some processes but not this one. Agitation was not allowed due to the fact that it would lead to the creation of fine particles that would get embedded in the filter cloth and make it unusable. Operators could use a water hose to fix cracks, but the automation system did not have these same facilities (e.g., eyes, arms, hands, and the ability to focus the water to any location on the cake). The detection of a cake was fairly simple for an operator. The operator simply opened the manhole and looked at the cake. Automating this step was difficult because detecting the cake proved to be almost impossible for the sensor and the controller. In fact, a reliable method for detecting cake was never found.

In addition to the issues involved with sensing, pumping, cake distribution, and estimating weight, there were also issues associated with batch automation. The building needed to be capable of processing multiple products with different sequencing requirements, but there were no preexisting procedures that lent themselves to defining equipment phases or recipes. Most of the procedures resembled statements like "the operator will determine how many tanks will be used and then open the inlet valves to the appropriate tanks." While this statement might be perfectly obvious to a human, it becomes a bit more difficult to explain to a phase. Additionally, each wash tank had to be treated as its own batch and given a unique batch ID. This meant that each tank had to be independent and execute its own recipe.

Once again, a batch executive needed to be implemented for each wash tank. However, in this building a new twist was encountered: the recipe manager had to have the ability to backup and restart the sequence at any previous step in the recipe. Although this is a standard capability of commercial batch executives, the functionality had to be incorporated into this particular custom system. This need to restart had to be dynamic and under the control of the operator. If after several washings of the HEOs the material was too acidic, then the material would have to be revacuumed, rewashed, and reslurried. Unfortunately, this was a requirement that was not entirely clear until actual product was being processed. Ultimately, the problem was solved in a reasonably simple manner without any real changes to the structure of the batch executive.

Since these buildings had never been automated in the past and since the engineering requirements were never defined in great detail, it became a significant challenge at startup to adapt the control system to meet the operational needs of the process while simultaneously meeting the engineering objectives. Some of the problems might (or might not) have been more readily solved by a commercial batch executive. However, other problems most assuredly would have challenged these packages. One operational requirement in particular was a significant challenge: receiving feed material from the Reaction building.

The distance from the Reaction building to the Wash building was about a quarter mile. When material was pumped from one building to another, there was always a quarter mile of material that had to be pushed through the pipeline into the next building. When production requirements were low, the material might sit for twelve hours in the pipeline and become quite cool, and these particular HEOs do not filter very well when cooled. Operators would typically distribute the cold material into three wash tanks by manipulating valves during a feed. In addition, one reactor batch was larger than the volume of the three tanks, and material could be pumped into the wash tanks faster than the vacuum system could pull material out of a tank. Thus an operator had to continually monitor levels and switch feed valves to keep the wash tanks from overflowing.

This problem presented an interesting challenge to the batch automation system, but it was ultimately solved by utilizing the notion of a controller-initiated hold and restart of each wash tank's Receive phase. Each tank was designed to request and then receive access to the feed pipeline before its feed valve would open. However, when the phase went to hold it would automatically release the resource, and then when it came out of hold it would have to reacquire the resource. This approach resulted in a solution that closely mimicked an operator's ad hoc procedure. The result was that the system performed on par with what experienced operators expected and provided a filtered and washed product that was very close to those produced by the best operators. This innovative use of the ISA-88.01 concept of Hold and Resume led to a higher degree of buy-in by the operators than would have otherwise been achieved. An example of the system with three tanks actively receiving material is shown in Figure 21.5. In the figure, tank 1 is held while tank 2 is receiving. Tank 3 is actively waiting for its turn to receive.

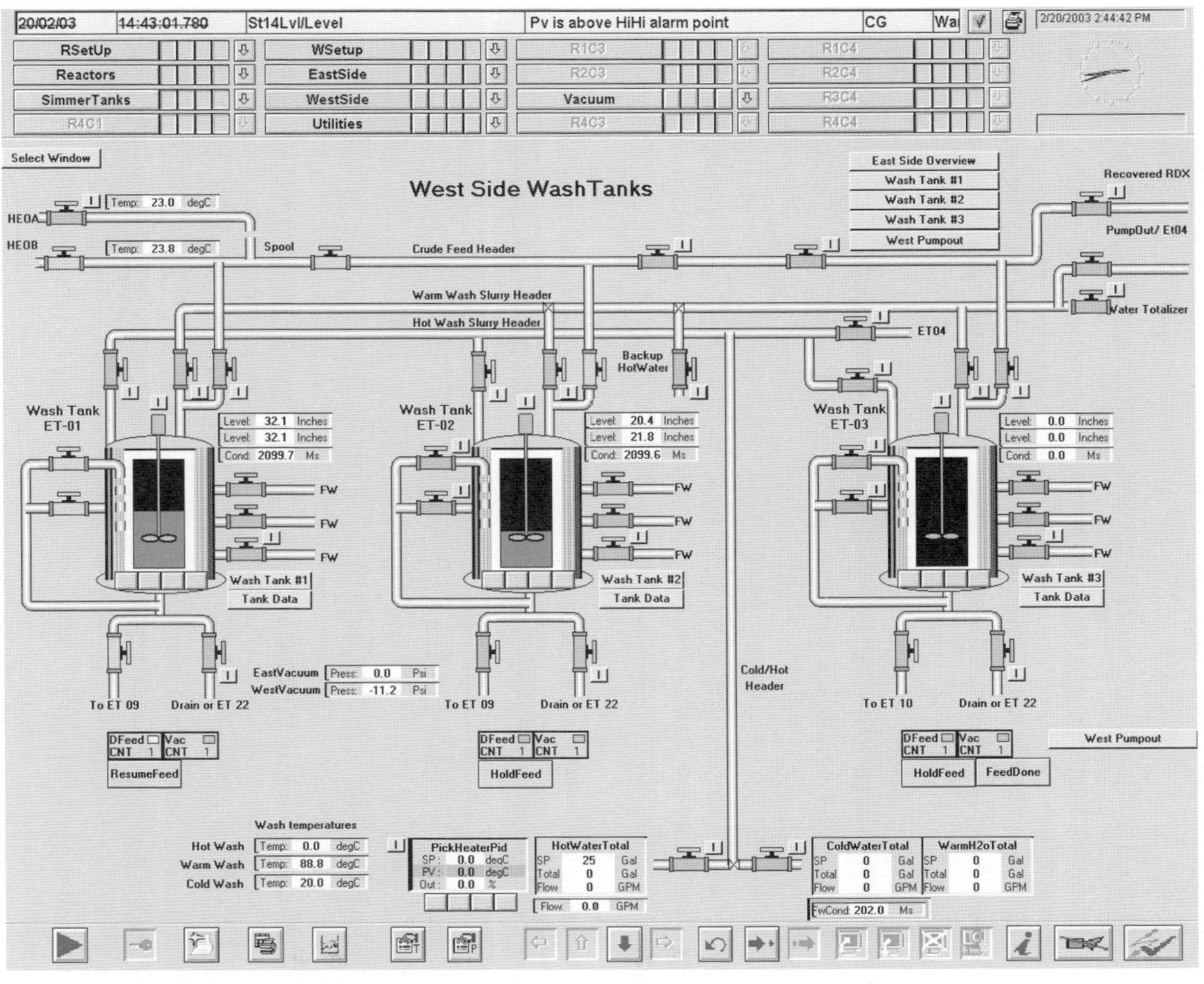

Figure 21.5. Overview of washing operations while receiving material.

Crystallization: No Two Systems Are Exactly the Same

The third stage in the manufacturing of HEOs is called Crystallization. In this stage the slurry from the Wash stage is introduced into a dissolver, mixed with solvents, heated, dropped into a still, and then boiled to produce crystals. This is the critical step in the manufacturing process. It differentiates the "raw" materials into the final products. There are over sixteen different products produced in the combined dissolvers and stills. There are two HEO types and three solvents. Each product has its own unique combination of solvents and heating profiles.

As with the previous building, a recipe manager and recipes were developed that could handle manufacturing a variety of different products. The primary goal of the automation system was to bring consistency and repeatability to the batch-wise manufacturing of these products. The focus of the development effort was to create phase logic and control strategies that led to the realization of this goal. The creation of a recipe manager to coordinate the phases was a secondary goal and was not necessarily considered that important to production management. In fact, the company was so comfortable with its manual operations in other buildings that it was over a year before the building was utilized in a production capacity.

The Crystallization buildings had been operated manually for over 60 years. The process of bringing automation to this manufacturing system was challenging. Many of the manufacturing steps were dependent on human judgment and decision making. In order to develop batch phases (or "steps" as some like to call them), it was first necessary to understand what operators actually did and then determine the criteria they employed in making decisions. Certainly the steps in the process were well known—definitely stepwise in nature and easier to quantify than in the case of the Wash stage. This made defining phases and a control recipe straightforward. However, it is one thing to be able to layout the problem, but it is another to actually replace human activity with computerized actions. In order to appreciate this, one must first know a bit about the process itself.

The slurry in the Wash building contained both HEOs and water. The actual quantity of water and HEOs were not known. The Standard Operating Procedure (SOP) defined a batch as containing X amount of HEOs. The SOP then specified a specific amount of solvent for each product. The SOP also called for a specific amount of energy to be applied for a specific amount of time. No consideration for variation in HEO quantity or solvent concentrations was specified in the SOP.

The SOP stated that the operator would add N pounds of solvent into the dissolver and then have the Wash building operator pump the HEO slurry into the dissolver. In order to accomplish this pumping, the building operator would manually set up the valve path, walk to the dissolver, and radio the Wash building operator to start pumping. The Wash building operator would then pump the material into the lines, add N gallons of water into the lines for cleaning, and then apply 100 pounds of air to blow the material through the pipeline into the dissolver. The operator in the dissolver building would wait until air was detected and then radio the Wash building operator that the material had arrived. This step was called the HEO charge step. The preaddition of solvent was called the solvent charge step. In the manufacturing of products, this approach to charging the dissolver with materials led to many variations in quality in the final products. One goal of the automation project was to reduce variations in product quality by automating these two steps.

The basic problem with these manual steps is the unknown ratio. The quantity of the HEOs was assumed to be N, but during water batches it was observed to vary by plus or minus 30%. Another problem was that the solvent addition was not always accurate. One solvent adsorbed water. Since the solvent was reused, there was often a buildup of water in the solvent. The other solvent floated on top of water, so any water that entered into the solvent tank sank to the bottom. Sometimes there would be 5000 pounds of water at the bottom of the tank. If the SOP called for 10,000 pounds of solvent, then only one half of the solvent required was added. These variations were significant. The manual process had no way of accounting for many of the variations because (1) there were no measurements and (2) there were too few experienced operators.

So the first order of business in development was to assume we would solve these problems. The process engineer specified sensors and techniques to overcome the measurement deficiencies. However, as with the Wash building, it was uncertain if the techniques developed to measure material concentrations would work. The next step was to define methods and computational techniques, build control elements, construct phases, and define recipes that utilized the equipment phases. With all this in place, we proceeded to water batching and started validating our solutions. Here is where the fun really began because many things did not work at first. The technologies employed for determining concentrations required significant adjustments. Operators did not always follow the SOPs, and the SOPs had not been updated to reflect actual practice. In addition, there are some things that just cannot be readily automated. Humans are necessary in many instances, and they cannot be replaced by automation. Nevertheless, automation can often be utilized to enhance operator performance. We learned this lesson repeatedly while trying to get the Crystallization buildings operational.

The first big problem was determining how to make the system detect when the slurry pumping was finished. This seemed simple enough in the laboratory: monitor the flow. However in practice it proved difficult. In the end we simply relied on the operator to tell us. This was somewhat disappointing, but it turned out to be the simplest and most cost-effective solution.

The next and most important problem was determining slurry concentration. This turned out to be very challenging. The approach that was taken was to monitor density and temperature to compute water in HEO concentrations and water in solvent concentrations. The same manifolds and sensors were used for both situations. Some methods had to be developed to differentiate one mixture from another. Pure water could easily be confused with being either a solvent or slurry.

Other complications of adding material resulted from the product recipes. Some recipes called for a specific quantity of ingredient. Others specified a maximum amount. Thus if the recipe called for the ADD HEO phase, then it had to be clear whether or not the setpoint was an end point or a maximum value. If the setpoint was a maximum value, then the system did not need to blow material back to the Wash building. If the setpoint was an end point, then when the setpoint value was reached and the feed valves needed to close, the operator needed to call the wash and filtration operator in order to stop the pumping and setup for a blowback, followed by a block and bleed operation. Many solutions presented themselves for these problems, but when a manual process is automated, many things operators do are taken for granted and are not thought of when the functional requirements are defined. Solving these issues may not be difficult in theory, but in the middle of a startup, they present interesting challenges.

In addition to just getting the buildings operational, there were also management issues. The plant management team wanted to get the projects completed and the books closed. This meant there was significant pressure to shorten the schedules. There was only one team—consisting of a process engineer and an automation engineer—designing and implementing the automation solutions. There was some parallelism in the projects, but many aspects had to be done serially. This led to problems, particularly with respect to the two Crystallization buildings.

There were two "identical" buildings that processed the materials out of the Wash building. From an engineering viewpoint, the process equipment in each building was different. However management did not perceive that there was a difference and continued to refer to the migration of the Building 1 program to Building 2 as a "copy." This meant that little time was allocated to Building 2 in the project schedule. Because of this, the project required that the Building 2 system get up and running with changes made on the fly as needed.

Since both Crystallization buildings were "identical," the project schedule did not include any time to address the actual building differences. This meant that in addition to the construction of a multiproduct and multistream batch solution, there had to be explicit structures embedded in the solution that addressed the different mechanisms in each of the buildings for receiving products, pumping solvents, dissolving HEOs, and many other things.

One example of the differences in these "identical" buildings was in the area of solvent recovery or decanting. In one of the buildings, there is a decant system in place that automatically handles the separation of solvent and water. In the other building, the decanting is performed within the dissolvers. In the first building, the dissolver's contents were dumped into the still, and then the dissolver was free to start a new batch. In the second building, after the contents of the dissolver were dumped, the dissolver had to continue and set itself up to act as a decant system. In the one unit, decanting consisted of three additional phases. In the other unit, the decant operation was nonexistent. One might argue that a recipe should merely call for a decant operation, and any differences should be managed by the underlying equipment phases. Regardless, something somewhere in the system must be designed and implemented to deal with the differences.

In our implementation of a recipe manager, we chose to handle this difference in the equipment via a configuration bit in the dissolver units and a construct in the data used to specify which phases to execute. We wanted to write one recipe for each product. We wanted to have the recipe execute in either building. We also wanted to be able to flip a switch to make the behavior in either building identical. We wanted a solution that was simple, clean, and easy to understand. The solution we chose met that requirement. When the unit recipe manager of the dissolver encountered the construct, it then used the configuration bit to decide whether to continue processing the data in the recipe or to quit. If the manager quit, then the recipe was finished and the dissolver was free. If the manager continued, then the data in the recipe table were read to determine which phase would be processed next. This worked very well, and the end customer was quite satisfied with the way the system operated.

Many of the other basic subsystems were also similar but different. The raw material delivery subsystem was another subsystem that exhibited the "similar but different" characteristics. These are illustrated by Figures 21.6 and 21.7. The system also had to cope with different raw material delivery methods. In one building, the raw material was dumped into a holding tank. The material in the holding tank was assigned a batch ID for tracking purposes. This holding tank could hold material for three or four downstream batches, and raw material would be pumped from the holding tank as needed. In the other building, material was received directly into the dissolver. There was no buffer for materials coming from the Wash stage. When the recipe called for the HEO A path (Figs. 21.6 and 21.7), in one building there was one set of actions. The same request in the other building resulted in another set of actions. Even though the recipe simply called for the system to Add HEO A, this did not change the fact that differences had to be dealt with by the phases in each building. The real problem was that no time

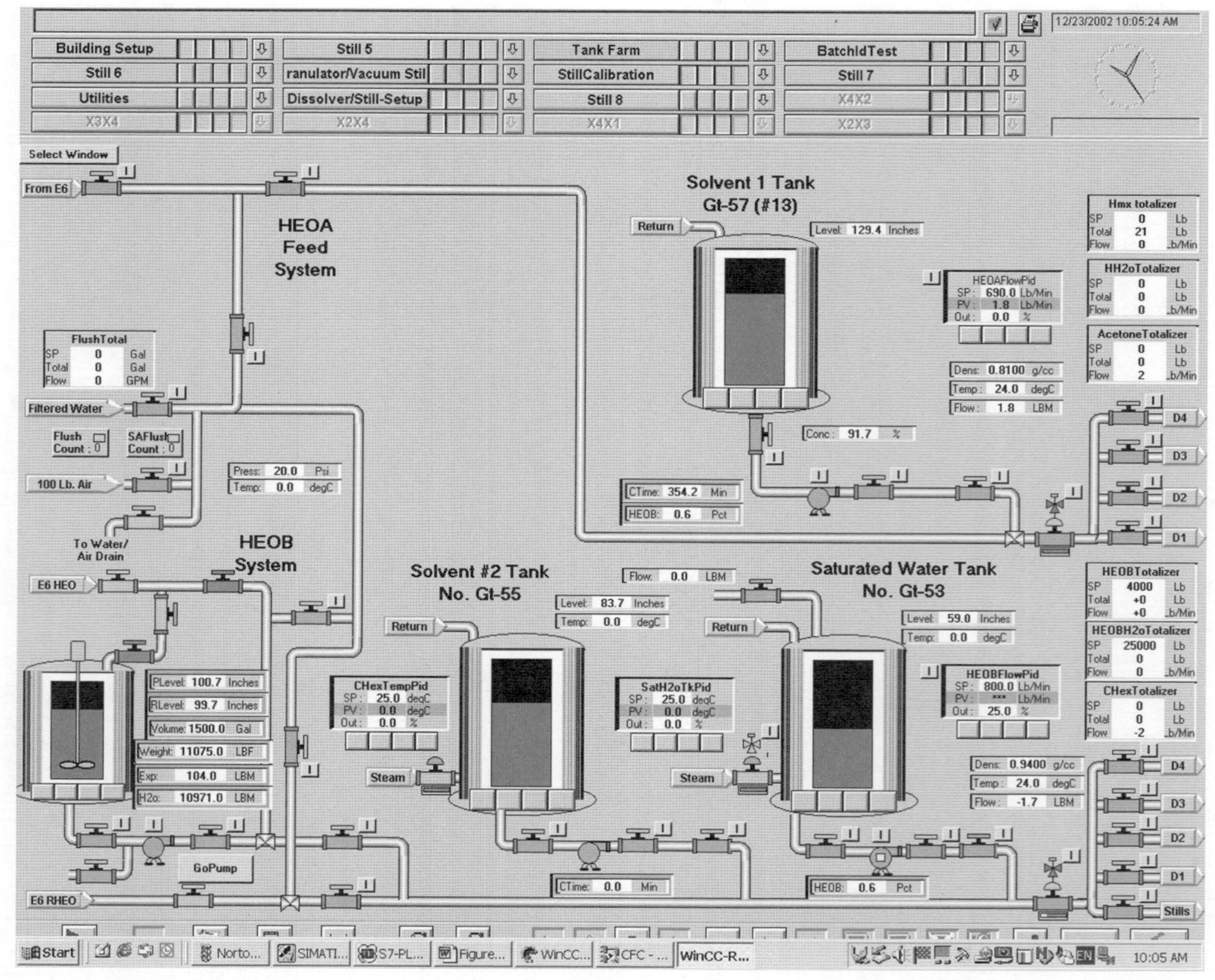

Figure 21.6.   Tank farm equipment for Building 1.

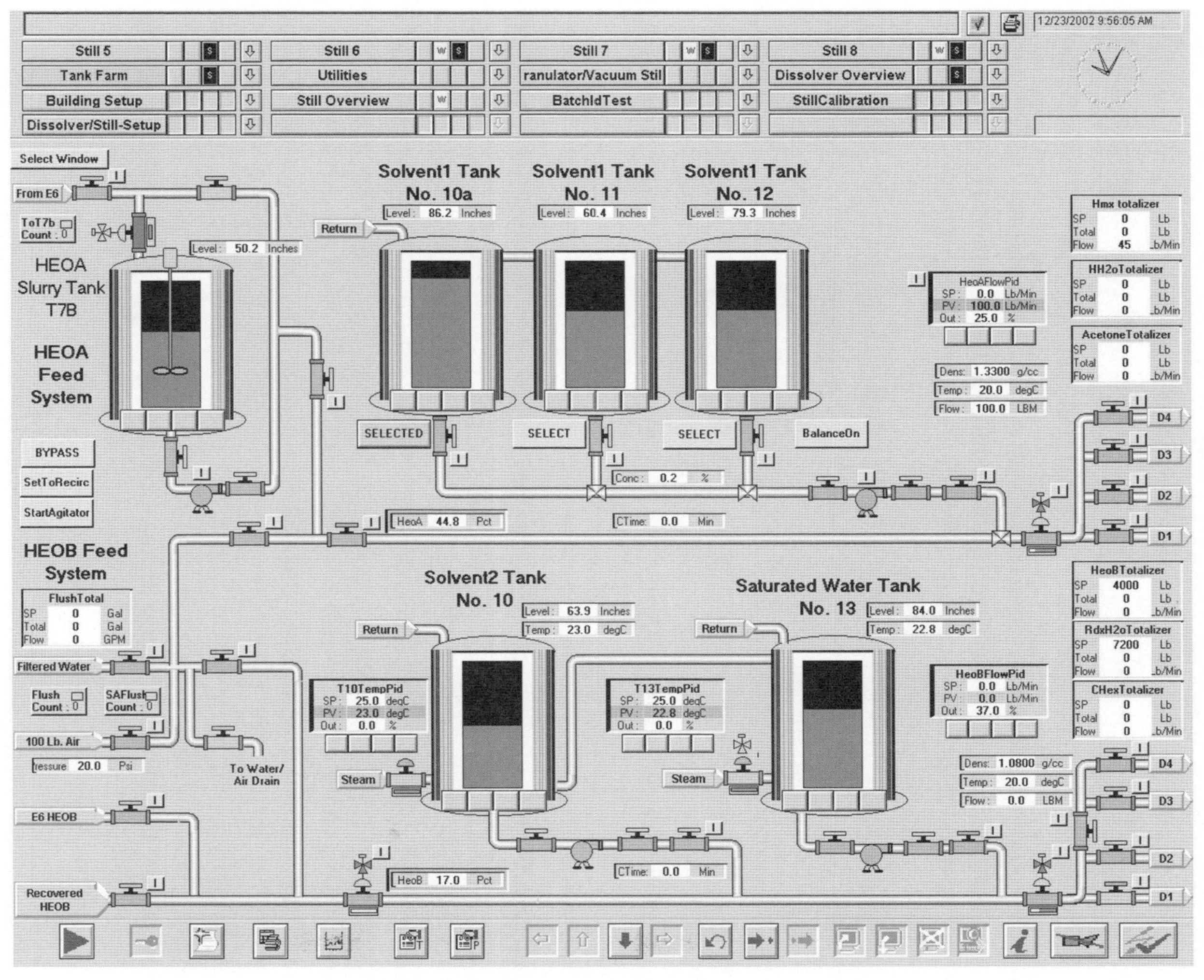

Figure 21.7. Tank farm equipment for Building 2.

was allocated in the schedule for addressing the differences. Changes were made during the water batching phase and even after the buildings were in production.

Another complication was that batches took more than one shift to complete, and operators were only available for one shift. Due to the nature of the materials, it was decided that nothing could be left active when the building was empty, including the recipe. In an ideal world this would have meant the following: at the end of the shift the batch would be placed into a safe state, and at the start of the next day's shift the batch would be resumed from its previous state. This corresponds to the ISA-88.01 Hold state and the Hold and Resume commands. This is not complex, and the recipe manager we constructed could have readily handled this solution. Hold states and Hold and Resume commands were available in the manager and the phases.

However, this is not what the customer required. The customer required that the batch be aborted. This did not mean "place the recipe in hold." This meant kill everything, cool the system down, lock down the equipment, leave the building, and lock the doors. In the morning we needed to resume processing at the exact point in the recipe starting from idle. The manager had to have the ability to reboot to the exact spot and continue processing. This added an interesting complication to an otherwise already complex problem. However, the solution proved to be most valuable and has found application for several other end users where conventional recipe managers were not utilized.

Summary

The ISA-88.01 batch models were successfully applied to a complex set of control system problems in the production of HEOs. This helped reduce the amount of off-specification products from 15 to 20% to less than 2%. It also improved plant throughput via more effective use of equipment and allowed for increased safety and reduced labor costs. These advantages did not come without some lessons learned, which include the following: (1) if you need batch flexibility, you have to buy it or build it, but there is still a cost; (2) sometimes the hardest problems are formalizing common sense and replacing human senses with sensors; and (3) there is no substitute for adequate design time—apparently insignificant differences from a macro viewpoint have significant consequences at the implementation viewpoint.

ISA-88.01 concepts were utilized whenever they were applicable. However, it is important to bear in mind that ISA-88.01 provides a set of concepts and does not define any specific solution. The ISA-88.01 models only define a framework to document and implement solutions. In the end, solutions must be developed that satisfy the customer's requirements. Waving a standard in the customer's face serves no purpose.

A data-driven batch executive provided the framework that allowed many of the operational requirements of the buildings to be met. Plant management just wanted operational solutions and really did not care to know the details of how that was achieved. However, pulling it all together and getting it to work in a consistent manner did call for a batch executive. This was one of the reasons that the batch manager was developed. It needed to be there regardless of whether management perceived it as a requirement. The use of phases built on the ISA-88.01 Phase definition also played an important role. They provided the glue that allowed solutions to be built that could be changed easily and expanded readily.

Material Transfers

William M. Hawkins
Owner
wmh@iaxs.net
HLQ Ltd., 10300 Colorado Road, Suite 342,
Bloomington, MN 55438, USA

Abstract

Batch processes make products in units that hold no more than one batch at a time. Something has to move gas, liquid, or solid materials in and out of the units. The materials move along a route that usually consists of pipes and valves, transfer panels, or hoses. Conveyors, wheelbarrows, or trucks may be used for solids.

Sources can be warehouses, storage tanks, or units. Destinations can be units, storage tanks, or warehouses. The route may include a filter. The flow may be measured. The route may have to be cleaned or sterilized before or after or before and after the transfer.

Another method for transfer is the "pipeless plant." The reactor unit is mobile and moves from one fixed station to another to add, process, or discharge material. That method will not be discussed in this chapter.

Modeling Transfers

In the general case shown in Figure 22.1, some kind of Source Control prevents material from leaving the source until the route is ready. If the route is not simple, Route Selection is necessary to select and prepare a route. Route Selection may use a transfer task for each route so that whatever requests the transfer does not have to know the details of the route. Transfer Control operates the transfer along the route. Destination Control prevents material from entering the destination before the destination is ready; this may be achieved via a person with a clipboard delaying a truck driver until everything is signed off.

The material may be "pulled" from the source by a request from the destination, or the source may "push" material to a destination. Workflow studies

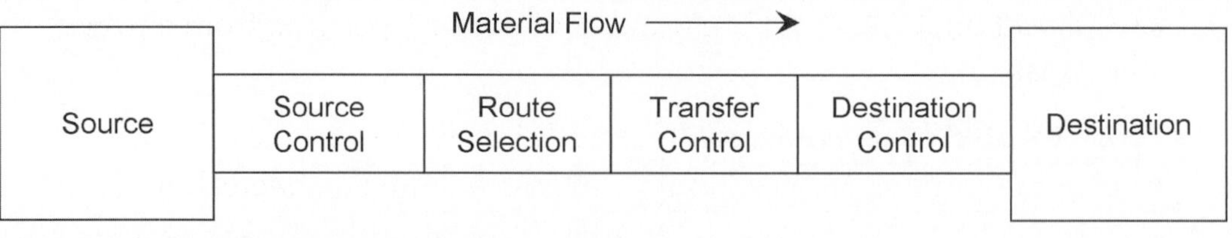

Figure 22.1.   Generic transfer model.

suggest that pulling is better than pushing, but sometimes the destination is storage that is selected by an operator or algorithm. The destination may not care what gets transferred to it, until it runs out of room. Some form of coordination control between source and destination is usually required. There is no point in starting a transfer to a destination that cannot hold all of it.

An ISA88 member wanted us to cease discussion of transfers because his units just notified batch management that they were done, and a separate transfer application determined where to put the batch. Unit recipes were unaware of transfers. This was a special case of a process that would have been continuous, had it not been for batch fermentation and the requirement to track individual batches. He did not want us to write something that would restrict what he was doing, and so ISA-88.01 does not discuss material transfers in any depth.

Properties of Transfers

The list of attributes for a transfer may include the following:

- *Material.* Any properties of the material that may affect the transfer
- *Previous material.* Any material properties that may be incompatible with material to be transferred
- *Source.* Location of the material to be transferred
- *Source state.* Ready or not ready, at a minimum
- *Destination.* Location of the receiver of the material
- *Destination state.* Ready or not ready, at a minimum
- *Specified route.* Used if Route Selection does not calculate a best route
- *Actual route.* The route taken, as a name or a list of intersections
- *State of route.* Used if the route must be cleaned or sterilized; may include stale
- *Stale timer.* Used to put a limit on time after sterilization

- *State of transfer.* Used if the transfer has states, perhaps from a transfer task

- *Time in transfer state.* Also used with a transfer task

- *Elapsed transfer time.* Used to time a transfer, if that is interesting

- *Total transferred.* Used if the amount transferred matters, as in a material addition

- *Fault conditions.* Describes any failures, such as valve position failures

Operator displays should show these properties. The historian will probably also be interested. Transfer tasks with states will be discussed after simple transfers are discussed in the following section.

The Simplest Transfer

A simple transfer task has a fixed source and destination so that no selections are required, as shown in Figure 22.2. The simplest task has one common valve with an inlet that belongs to one tank and an outlet that belongs to another tank. When the tanks are units, the common valve is the dump valve of one unit and an inlet valve for another unit. Two units cannot own one valve, to avoid conflicting commands.

The common valve problem has existed from the early days of batch control. Many solutions have been tried, and all were dependent on the capability of the control system being used. Figure 22.3 shows the common valve between the upper and lower unit. The unit borders split the valve, which is not possible because the valve has only one command input.

ISA-88.01 might lead you in the direction of calling the valve "common equipment." This would mean that the Control Module (CM) must handle multiple requests for acquisition. The units on either side of the valve would have to acquire it, but nothing else can acquire the valve, making it different from shared-use equipment in that only two, not one, acquisitions must occur. This seems to be an unnecessary complication for a single valve in a fixed position.

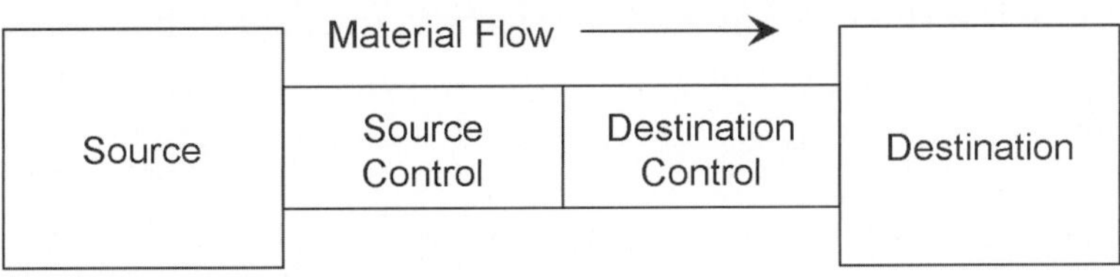

Figure 22.2.   Simple transfer model.

Another possibility is to have the source own the valve. The destination must somehow request the source to open the valve and to close it when it is done. If communication is interrupted, the valve stays in its last position. This can be fixed by using two valves in series, but that's expensive.

Modern control systems allow the discrete state device control function to have an interlock input. This allows two different sources of commands to control the valve. Let the source unit own the valve and the open/close command for the valve. Let the destination unit own the interlock. The source can command the valve to open at any time, but the valve will not open unless the destination has enabled the interlock.

Figure 22.4 shows the common valve completely inside the border of the upper unit. The CM for the interlock switch is completely inside the border of the lower unit. The configuration in Figure 22.4 lets either unit close the valve and stop the transfer, but only the source can open the valve. If your design philosophy says that a unit owns all its inlet valves, then you can turn the configuration around. Let the source control the interlock and the destination control the valve. Either way, both units need to know the actual valve position.

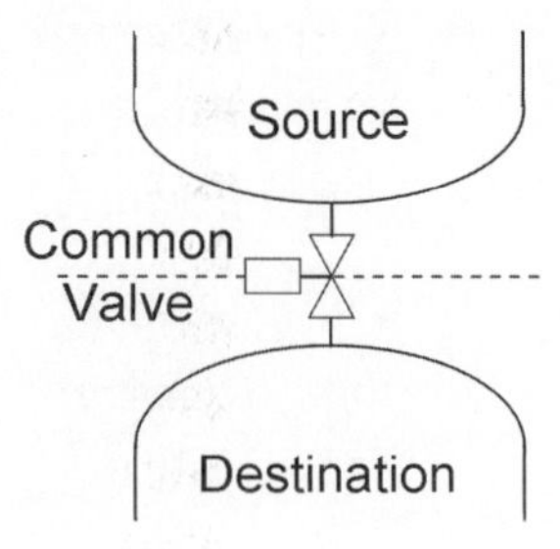

Figure 22.3. Transfer with a single valve.

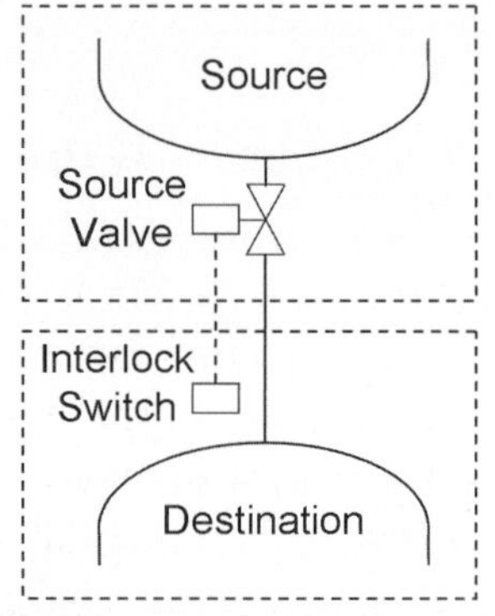

Figure 22.4. Joint control with an interlock.

Simple but Not That Simple

The situation is different if more than two units are involved (Fig. 22.5). Configuration 1 shows that the upper unit is a single source to two or more destinations. The upper unit owns the valves and makes the selection. The lower units can prevent flow with their interlock switches. Configuration 2 shows the case for a common destination. In either case, valves and interlocks could be swapped vertically if that better suits your purpose.

Where there is a choice of methods to handle transfer valves, pick one and be consistent in applying the method. People make mistakes when different methods are used (such as when vendors use different panel arrangements for appliance controls).

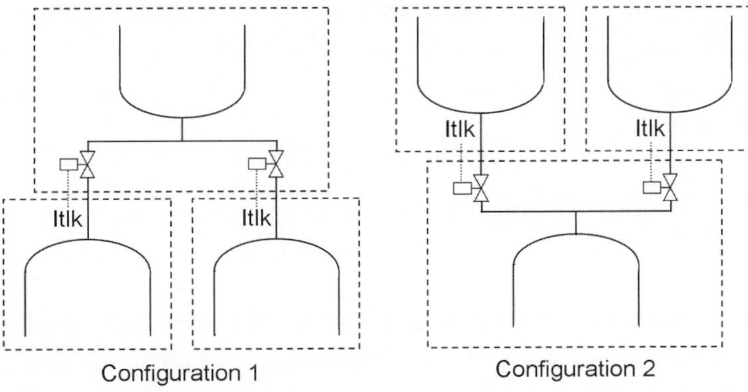

Figure 22.5. Multiple units with single valves.

Header Modules

Figure 22.6 shows multiple sources and destinations. Now you need a separate header module that can be acquired. One of the source units acquires the header module and selects a destination. You may still want each source and destination to have an interlock on their valve from the header. It is not practical to have the destination own the header valve because the acquirer will set the path.

The type of header module depends on whether or not the recipe can interact with it. If the source unit simply acquires the module and sets a setpoint to

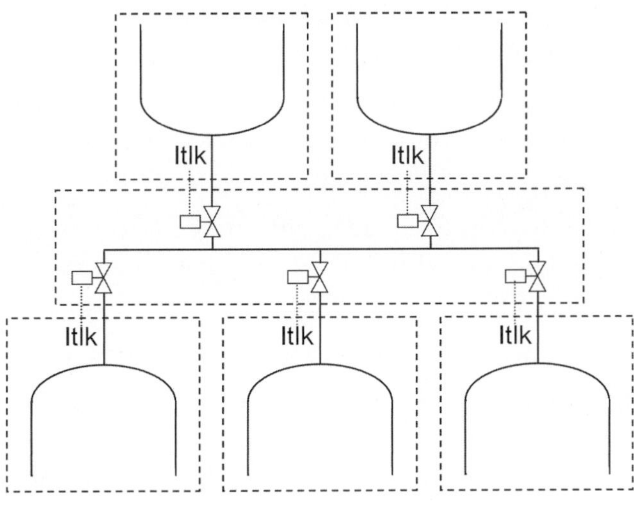

Figure 22.6. Simple transfer with a header.

determine the destination, then the module is a CM. This is true even if it runs a complex cleaning sequence, providing the sequence does not change with the recipe. If it does, then you have an Equipment Module (EM) that runs a phase that is responsive to the recipe.

Transfer Tasks

The final complication is a multiple simultaneous transfer situation that requires many valves. Imagine the holding tank facility in a large brewery with one-hundred tanks. About 3000 Inputs/Outputs (I/O) are required to do simultaneous transfers and separate the product from the cleaning solution. All 3000 must be tested 120 times per second (less for 50 Hz mains). Each source must ask for a route and acquire it, up to the capacity of the header network. The algorithm to create routes through the maze can be quite interesting, especially if each route requires a pump.

A transfer task may have the following states:

- *Idle.* Be ready to perform a transfer, possibly counting down the stale timer for sterilization.

- *Close.* Confirm that there are no open cross-connections with other routes; fail if this is not true.

- *Open.* Open and confirm anything that would block the route; fail if this is not possible.

- *Prepare.* If required, clean or sterilize the route; skip this step if the purpose of the transfer is to sterilize the source, destination, and the route between them.

- *Start.* Request or wait for destination control to open; when confirmed, request or wait for source control to open, and after this is confirmed, change the state to transfer in progress.

- *Transfer in progress.* Monitor the route connections and fail the transfer on any problem; possibly fail if the transfer has exceeded its time limit.

- *Stopping transfer.* Request or wait for source control to close.

- *Drain.* If required, allow time for the route to drain into the destination; then request or wait for destination control to close.

- *Clean.* If required, perform a cleaning procedure, or tell a subordinate task to do it.

- *Sterilize.* If required, perform a sterilization procedure, or tell a subordinate task to do it.

- *Done.* Do any reporting required, and reset the stale timer, if any, to a given time; if necessary, notify coordination control that the transfer is done and that the transfer task may be released.

- *Hold.* Stop the transfer but leave the route occupied.

- *Resuming.* Restart the transfer using the same route.

- *Abort.* Do so if necessary and feasible; otherwise just stop.

- *Failed.* Close the route and stop any energy source; await manual reset.

A transfer task may be allocated or reserved before being acquired. The task may need to select the path to the first or best available destination or fail to find a path. The complexity of a transfer task grows with the number of sources and destinations that share a transfer network.

Bottom Line

Transfer tasks have to run in the control system being used for the batch processes until they reach a level of complexity that justifies a separate system. This is a problem because there is no "models and terminology" standard for transfer tasks. All sorts of terms and work-arounds are used, as in the days before ISA-88.01.

Index